Hiroba : All about "Public spaces" in Japan

Editorial supervision by Kengo Kuma, Hidenobu Jinnai
監修　隈　研吾
　　　陣内秀信

Photographs by Tomoyuki Suzuki
写真　鈴木知之

Tankosha
淡交社

Hiroba: All about "Public spaces" in Japan

Editorial supervision by Kengo Kuma, Hidenobu Jinnai
監修　隈　研吾
　　　陣内秀信

Photographs by Tomoyuki Suzuki
写真　鈴木知之

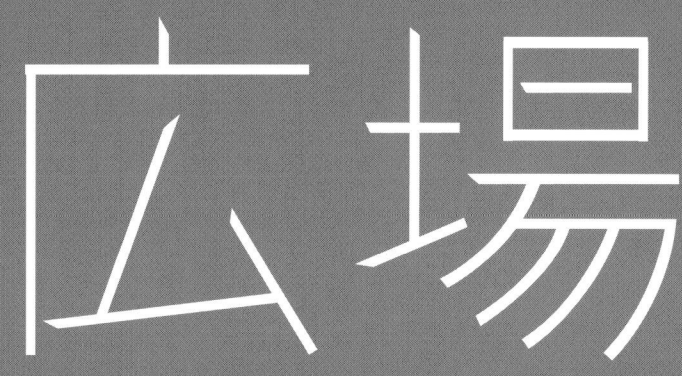

Tankosha
淡交社

Hiroba: All about "Public spaces" in Japan

Published in 2015 by Tankosha Publishing Co.,Ltd.
Copyright 2015, Tankosha Publishing Co., Ltd, All Rights Reserved.

Printed in Japan
ISBN978-4-473-04014-5

広場

Hiroba:
All about "Public spaces" in Japan

Editorial supervision by Kengo Kuma, Hidenobu Jinnai
監修　隈　研吾
　　　陣内秀信

Photographs by Tomoyuki Suzuki
写真　鈴木知之

Tankosha
淡交社

目次
Contents

実体への回帰
Back to Urban Reality
隈 研吾　Kengo Kuma　*6*

隈研吾 建築作品にみる「広場」
"Public Spaces" in the works of Kengo Kuma

アオーレ長岡　　　Nagaoka City Hall Aore　*18*
la kagu　*22*
木挽町広場（歌舞伎座）GINZA KABUKIZA, Kobikicho Plaza　*26*
マルセイユ現代美術センター　　FRAC Marseille　*30*
ブザンソン芸術文化センター　　Besançon Art Center and Cité de la Musique　*32*

日本独自の広場、その多様性の検証　Diversity of Unique Japanese Public Spaces
陣内秀信　Hidenobu Jinnai　*34*

日本の広場　事例集　*Public Spaces in Japan*

金沢21世紀美術館　21st Century Museum of Contemporary Art, Kanazawa　*54*
熊本駅東口駅前広場（暫定形）　Kumamoto Station east exit station square　*56*
東急プラザ 表参道原宿「おもはらの森」
　　　　　　　　　TOKYU PLAZA OMOTESANDO HARAJUKU, "OMOHARA-NO-MORI"　*58*
録 museum　Roku museum&cafe　*62*
東京国際空港第2旅客ターミナルビル　Tokyo International Airport Terminal 2, "UPPER DECK TOKYO"　*64*
りくカフェ　　Rikucafe　*68*
KOIL（柏の葉オープンイノベーションラボ）イノベーションフロア
　　　　　　　　　　　　　　Kashiwa-no-ha Open Innovation Lab., Innovation Floor　*70*
姫路駅北駅前広場　Himeji station North square　*74*
SHIBAURA HOUSE　*78*
木屋旅館　　KIYA RYOKAN　*82*
六本木ヒルズ 66 プラザ　Roppongi Hills 66 Plaza　*84*
仏生山温泉（仏生山まちぐるみ旅館）Busshozan-onsen (Busshozan-machigurumi-hotel)　*86*
代官山ヒルサイドテラス　HILLSIDE TERRACE　　　*90*

代官山 蔦屋書店　DAIKANYAMA T-SITE　92

マーチエキュート神田万世橋　mAAch ecute KANDA MANSEIBASHI　94

東北自動車道 羽生パーキングエリア（上り線）　The Tohoku Expressway Hanyu Parking Area (Up Line)　96

サンストリート亀戸　Sun Street KAMEIDO　98

権堂パブリックスペース OPEN　Gondo Public space OPEN　100

アーツ千代田 3331　3331 Arts Chiyoda　102

SCAI THE BATHHOUSE　104

新宿マルイ 本館　屋上庭園「Q-COURT」　Shinjuku Marui Main Building, rooftop garden "Q-COURT"　106

目黒天空庭園　Meguro Sky Garden　108

丸の内ブリックスクエア　三菱一号館広場　Marunouchi BRICK SQUARE　112

虎ノ門ヒルズ　オーバル広場　Toranomon Hills OVAL PLAZA　114

都市の余白　A blank space in the city center　115

興福寺　薪御能（薪能）　The Kohfukuji Temple, Takigi-Onoh (Takigi-Noh)　116

旧金毘羅大芝居（金丸座）　The Former Kompira Oshibai (Kanamaruza)　117

門前町（善光寺）Cathedral town (Zenkoji)　118

参道（金刀比羅宮）　The approach to Shrine (Kotohira-gu)　119

鬼子母神　手創り市 Kishimojin, Marche　120

花園神社　酉の市　Hanazono Shrine, "Tori-no-ichi"　122

巣鴨地蔵通り商店街（「とげぬき地蔵」）　Sugamo Jizo Street Mall ("Togenuki Jizo")　124

阿佐ヶ谷駅前広場　Asagaya station square　126

ポンテ広場　PONTE SQUARE　128

歩行者天国（銀座）　Pedestrian mall (Ginza)　130

お台場海浜公園　Odaiba Marine Park　132

カナルカフェ　CANAL CAFE　134

お鷹の道　Otaka-no-michi　136

隅田川テラス　Sumidagawa Terrace　137

ネオ屋台村® 有楽町東京国際フォーラム村　NEO STALL VILLAGE®　138

太陽のマルシェ　Taiyo-no-marche　140

橋詰広場（日本橋）Bridgehead plaza (Nihombashi)　142

新たなパブリックスペースの復権　*Restoration of New Public Spaces*
小野寺 康　Yasushi Onodera　145

空間を超えた広場の在り方　*Hiroba beyond Space*
永山祐子　Yuko Nagayama　158

掲載事例　所在地一覧　Location List　164

退屈な付属物としての過去の広場

　広場がリアリティを取り戻しつつある。リアリティを取り戻すとは、広場の復活に他ならない。人間が広場を取り戻しつつあるということである。

　広場の歴史的変遷について考える時、いつも想起するのは、建築家磯崎新（1931〜）が、日本の高度成長期の只中（1968年）、『日本の都市空間』（彰国社）の巻頭に掲げた、都市デザインの4段階説である。

　1) 実体論的段階、2) 機能論的段階、3) 構造論的段階、4) 象徴論的段階の4段階を経て都市デザインは変化するというのが磯崎の説であった。磯崎が下敷きとしたのは、新カント派の哲学者カッシラー（1874〜1945）が十八番にしていた4段階説である。

　この説が説得力を持ったのは、これが人間の頭の中でよく起こる4ステップの進化であるだけでなく、20世紀の経済・社会的状況の進化と、見事にシンクロしていたせいである。すなわち現実の広場のデザインが、19世紀から20世紀にかけてこう進化したからである。

　順番にたどってみよう。まず、産業革命以前の都市は、実体として認識され、デザインされてきた。すなわち 1) 実体論的段階にあった。人間がそこに生活するわけであるから、都市を実体としてデザインするのは当然である。住宅やレストランをデザインする時と同様に、あるいはそこに置く小さな家具や調度を選んだり、並べたりするのと同じような、きめの細かい方法を用いて、都市

実体への回帰　　隈 研吾

Back to Urban Reality　　Kengo Kuma

Trivialised public spaces of the past

Public space is making a comeback. People are reclaiming the reality of shared communal space and bringing it back to life.

When reflecting on the history of public space, I always think of Arata Isozaki (b.1931) and his four stages of urban design cited at the beginning of *Japanese Urban Space (Nippon no Toshi Kukan,* Shokokusha, 1968) based on the stock 'four stage' phenomenology espoused by Neo-Kantian philosopher Ernst Cassirer (1874-1945):

　1. Realism　2. Functionalism
　3. Structuralism　4. Symbolism

Particularly convincing is his thesis that these four stages commonly observed in the evolution of human consciousness might also be applied to 19th to 20th century economic and social advancements as seen in the physical design of public space.

Taking this step by step: the city prior to the Industrial Revolution was regarded and designed as a physical reality, i.e. the realist stage. It was obvious to city-dwellers that they had to physically design their city. Almost as if decorating a home or restaurant or arranging furnishings and accoutrements, they designed large plazas and even whole cities in exacting minutiae.

Came the Industrial Revolution, however, mass industrialisation radically changed urban conditions to where such a detailed approach could no longer cope and even functionalist partitions did not suffice. More than just dividing up the existing city, it became necessary to expand vastly outward. A more comprehensive structuralist approach, such as championed by Kenzo Tange (1913-2005) in his *Tokyo Bay Plan 1960,*[1] was needed to control that expansion. Emerging economic powers frequently replicated images of violently extended super-structured linear cit-

も広場という大ものもデザインされてきたのである。

しかし、産業革命以降の工業化の波の中で、都市の状況は激変した。小さな方法、実体論的方法では対応できないような非常事態が発生したのである。

まず、工場と住宅地を隔離する必要が生じた。工場、住宅といった機能ごとに、都市を大きく分割して、混じりあわないようにする必要が生じたのである。

しかも、2）機能論的分割だけでは不十分であった。既製都市を整理し、分割するだけではなく、外側に向かって大きく拡張する必要が生じたのである。拡張をコントロールするためには、第3ステップである構造論的なアプローチが必須であった。丹下健三（1913～2005）の「東京計画1960」が構造論のチャンピオンである。強い構造的軸線に沿って、リニアに拡張するこんな暴力的な都市の絵が、経済の高度成長期を迎えた国々で、頻繁に描かれたのである。その構造論的都市デザインの中で、広場はしばしば軸に付属する、抽象的で実体のないオープンスペースとして、軽視された。

さらに、20世紀後半以降、新しい状況が出現した。グローバル資本主義、すなわち国境を越えて流動する金融資本が経済を先導する時代が来たのである。象徴性の強い、特異な形態を持った超高層ビルが、この時代の主役であった。投資先を求めて、世界を漂う資本による新しい都市ゲームがはじまったのである。シンボリックな超高層ビルが発する強い磁力が経済をドライブし、都市の

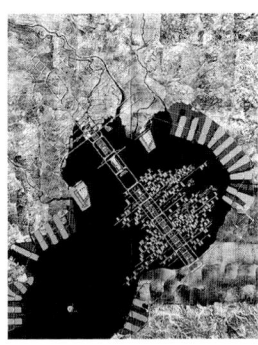

「東京計画1960」
1961年に丹下健三が発表した東京の都市構造への改革案。人口1000万人を超え、求心型・放射状の「閉じた」都市構造であった従来の東京を、都心から東京湾を横断し千葉県木更津方面へと至る「開いた」都市に変える海上都市構想を大胆に提案、当時各界に多大な影響を与えた。

Unveiled in 1961, Kenzo Tange's master plan for revolutionising Tokyo envisioned expanding the existing "closed" centripetal radial urban structure with its over-10m population by means of a floating "open" structure spanning eastward across Tokyo Bay from the city centre toward Kisarazu in Chiba. This bold new plan influenced many other fields at the time.

ies, often incorporating abstract and undefined open areas within their axes as if in passing gesture to public space.

Then from the late 20th century, the advent of yet newer urban conditions spearheaded by global capitalism and transnational movement of finances led to a predominance of high-rise corporate buildings branded in various emblematic forms. In the new urban game of seeking where to invest and how to attract floating capital, symbolic skyscrapers exerted a strong magnetic force to drive economies while imparting a decisive look to cities. This magnetism demanded yet another reshuffling of the urban core, reducing public space to the accessorised banality of shopping concourses paired with fashion boutiques around the base of overshadowing towers.

In sum, first functionalist façades replaced the intricate reality of the pre-industrial city, then urban redevelopment along structuralist lines maximised the flow of cars and materials, until finally principles of financial fluidity utterly divorced from physical lines and planes dissolved urban space into concentrations of data points.

Thus, while cities may appear to have blossomed spectacularly, the most overlooked, rejected element in the whole process has been people themselves. The four-stage evolution has only succeeded in dismantling and destroying the city.

Rehumanising today's cities

Very well, what does the future hold for our cities? Or rather, which direction should we take them? Beyond his four stages, Isozaki posited an 'invisible city' wafting around us like smoke, a 'black post-history' of connectivity where unseen computer networks would ever more totally dismantle our physical urban reality.

Yet looking over the examples in this book of new Jap-

イメージを決定する時代がはじまったのである。磁力を発する点によって、都市は再編成された。その時、広場は塔の磁力を高めるための、退屈な付属物になった。広場は一群のブランドショップとセットになって、塔の足元を飾る装飾にまで堕ちたのである。

総括すれば、産業革命以前の実体としての都市に代わって、まず機能論的な面としての都市が出現し、続いて構造論的な線によって物と車のフローが再編成された。ついには、面とも線とも無関係に自由に流動する、マネーという存在の原理に従って、都市は点の集積へと解体されていったのである。

都市は一見、派手が花が咲きほこるがごとき様相を呈するようになった。しかし、このプロセスの中で、最も無視され、都市から排除されたのは、人間そのものであった。4段階の進化とは、都市の解体、都市の破壊に他ならなかったのである。

都市が「人間に取り戻された」現代

では、この先にどのような事態が都市を待っているのだろうか。あるいは、われわれは、都市をどのような方向へと持っていくべきだろうか。

磯崎新は、この4段階の先に来るのは「見えない都市」であるといって、人々を煙に巻いた。コンピューターが、様々な流れを不可視化し、都市は、さらに徹底的に解体されるであろうという、磯崎が得意とする一種のブラックヒストリーである。

anese public spaces — *hiroba*, literally 'wide open areas' — we find nothing of Isozaki's prophet-of-doom despair and finality, but rather a revival of bright, comforting physical reality. An unmistakeable human presence pervades, full of ingenious new schemes for reconnecting people with urban space. There's a welcome sense of things having come full circle and returned to the realist stage. Granted small spaces still represent only tiny fractions of gigantic cities, yet something here transcends those four stages; something beyond the violence of globalism is being born.

What are the underlying circumstances? For one thing, we are reaching a saturation point with global capitalism itself. Towers and plazas designed for immediate visual impact are all too quickly neglected by man and media alike, tossed on the scrap heap, and replaced by new towers that fare no better. People have not only tired of this vicious circle, even financing bodies are learning to beware of the fragile fate of these towers.

New currents are breezing onto the scene. Completely counter to Isozaki's destructive computer technology, we are seeing people reconnect with the city. In today's world of unlimited information access it almost seems as if every street, facade and pavement were digitised online, yet this itself has prompted a decisive reversal. There exist things that cannot be experienced virtually via the internet: the singular body-feel reality of walking around on bare ground, the sensory overlap of textures, sounds and smells of the city, physical sensations that simply cannot be uploaded or transmitted. We've noticed their absence from our digital daze and we're starting to miss them. The more everything goes digital, the more we're rediscovering, in ways no one would have expected, just how vital real things are. Computer technology has paradoxically breathed new life into the good old physical city.

しかし、本書に収められた日本の新しい広場を眺めると、磯崎が予言した絶望的、終焉的な状況とはまったく異なる様相を持った、明るくて、心が休まるような実体が復活しつつあるのを感じるのである。ここには間違いなく人間がいる。人間と都市とを再びつなぎあわせるための、様々な新しい工夫が輝いている。世界が一巡して、再び実体論的段階に戻ったような、何かほっとさせるような印象があるのである。まだ巨大な都市の中では断片と呼ぶしかないような小さい空間ではあるが、ここには4段階説を超えた何か、グローバリズムの暴力の先の何かが、生まれつつあるのである。

この裏にはどのような事情があるのだろうか。ひとつはグローバル資本主義自体の飽和という状況である。即効性のヴィジュアル・インパクトを目的としてデザインされた塔や広場が、いかに短期間に人々に飽きられ、メディアに忘れられ、巨大な廃棄物となりはてるか。それにとって代わる次なる「塔」も、一瞬後には同じ運命をたどるしかない。次々に塔の廃棄物として堆積していく。人々がその循環に辟易しただけではなく、金融資本自体が、塔というものの宿命ともろさとを学習しはじめたのである。

さらにもうひとつの新しい風が吹きはじめた。コンピューターテクノロジーが、磯崎が指摘したような破壊的方向とは全く逆の形で、都市と人間を再び結びつけはじめたのである。

情報テクノロジーは、ネットに載せられた、あらゆる情報へのアクセスを容易にした。世界のすべては、電子情報化されて、ネット上を漂ってい

Familiarity from new materials

Looking over these *hiroba* ideas, we can also feel other winds of change. Digital technology isn't the only thing that has evolved; there have been great advances in material technologies as well.

Back before industrialised society and the automotive age, Western cities were built of durable stone and brick and mortar, most of which remained useful when urban planners and builders sought robust materials capable of withstanding traffic exhaust and vibrations. Little if any retooling or renovation was needed.

Whereas in Japanese cities, public spaces were basically made of wood and paper, delicate materials never designed to withstand modern strains and stresses. The moment cars entered the picture everything had to be redone from scratch. The entire history of Japan's refined urban spaces was simply discarded as our urban planners imported Western architecture and public space design: the brick and mortar buildings, stone-faced squares and metalled avenues of the early modern Meiji Era (1868-1911). Then after World War II, following the four-stage programme, concrete became the main building material so as to accommodate Le Corbusier-influenced structuralist schemes of flow and expansion. This also meant violent interventions to pave Western-style public spaces completely unlike the human-scale *hiroba* we Japanese had traditionally fashioned of 'warm' renewable materials.

These developments were by no means limited to Japan; 'cold' concrete and steel trampled and destroyed public spaces the world over. Sadly, given the stringent conditions dictated by cars, industrialisation and fire hazard, there was no other option at the time.

Not until things began to change after the end of the

る。ストリートのすべてが、ファサードのすべてが、地表のすべてがネットにスキャンされたかにみえた。しかしまさにその時、決定的な逆転がもたらされたのである。

　どうしてもネットに載せることのできないものが見つかったのである。

　このひとつだけの自分の身体を使って、地面というごつごつとした実体の上を歩き回るというリアルな体験。都市のテクスチャーと、音と、匂いとを自分の身体に重ねあわせること。それだけは電子化できず、ネットに載せようがないことに、われわれは気づいたのである。ネット化の完成によって、逆にリアルなものの重要性が、誰もが予想できなかったような形で、再発見されるに至った。われわれはネットのおかげで実体の価値と意味とを再発見したのである。コンピューターテクノロジーが、逆説的に、あのなつかしい実体的都市をもう一度呼び戻したのである。

新素材がもたらした「なつかしさ」

　さらにこれらの広場をじっくりと眺めると、もうひとつの追い風が吹いていることも感じられる。ネットテクノロジーだけが進化したわけではない。素材にまつわる技術も大きく進化したのである。

　かつては、都市のパブリックスペースには、車の排気ガスにも振動にも耐えるロバスト（強靭）な素材が求められた。ヨーロッパの広場は、車の時代がやって来る前から、石やコンクリートと

20th century and architectural technologies entered yet another stage. For now we are seeing a whole range of new techniques for fireproofing and strengthening natural materials previously regarded as too brittle or flammable or prone to decay. Popular demand for natural materials has spurred major technological breakthroughs in architecture and public environs.

Among these, new techniques for treating wood have been especially dramatic, making non-flammable wood the new material of choice for finishes — or even for structural members in larger urban buildings. This, of course, has turned public space design around and enabled the gentle refinements of wood to grace our cities again.

Likewise, there have been technological breakthroughs in glass, long a primary building material in industrialised countries worldwide where people happily replaced dark heavy walls with clear windows. Since the late 20th century such innovations as plastic-augmented sizes, opacity-control glazing and diverse surface treatments have suddenly generated a wealth of new possibilities in glass. Unlike anything seen in modernist architecture, glass can now deliver near-infinite transparency or aesthetic subtleties to rival the most intricate expressions in traditional *shoji* papered windows.

Without a doubt, these innovations have dealt us a strong hand toward reclaiming human qualities in public space. Advancements in structural engineering have also helped considerably, allowing us to use finer wood posts in place of solid steel columns or wood together with glass for a more intimate feel. Indeed, we are seeing a return to familiar vocabularies, forms and textures.

Thanks to these new technologies and new materials, Japanese architects and designers are now taking the lead in regenerating more 'human' public space, if for no other

いった、耐久性のある素材で作られていたので、車の時代、工業化社会がやってきても、そのままでよかった。作り直し、デザインし直す必要がなかった。

しかし、日本では、都市のパブリックスペースも、基本的に、木や紙のようなやわらかくて繊細な素材で作られていた。車の時代、工業化社会に耐えられる素材、デザインではなかったのである。車がやってきた途端に、全面的な作り直し、ゼロからのスタートが必要だった。それまで積み上げてきた日本の繊細な都市空間を、チャラにしなければならなかったのである。

とりあえず、ヨーロッパの広場のデザインが輸入された。明治時代には、石やレンガで広場や大通りがデザインされた。第二次世界大戦後には、コンクリートを主な材料として、コルビュジエ流の、先述の4段階説に従えば、構造論的なフローと拡張とを重視する、暴力的なパブリックスペースが、多数作られることになったのである。

このような西洋輸入のパブリックスペースは、日本人が、木や紙のようなやさしく、ヒューマンな材料を用いて作りあげてきたパブリックスペースとは全く異質であった。コンクリートと鉄が、日本だけではなく、世界のあらゆるパブリックスペースを蹂躙し、破壊した。しかし、車、工業化、あるいは大災害の可能性といった、タフな諸条件をクリアするのは、とりあえず、そのオプションしかなかったのである。

しかし20世紀以降、建築技術は新たな段階を迎えた。

壊れやすく、燃えやすく、耐久性にも問題があるとされていた自然素材を、不燃化したり、腐り

reason than Japanese urban space originally had such refined and delicate qualities. We only forfeited them temporarily due to Westernised industrialisation; we still retain memories of our heritage of public space. The *hiroba* designs in this book prove it: for all their new sensibilities and techniques, they feel somehow very familiar.

Reconstructing a familiar new physical reality

A major return to physical reality is sweeping cities the world over. Now that the storms of industrialisation have passed, rapid economic growth has subsided, and even cars are slowly retreating from the cities, Japan has a great role to play within these new circumstances.

The West gave us many different real examples of public space. Isozaki once cited Austrian architect Camillo Sitte's (1843-1903) *Der Städtebau nach seinen künstlerischen Grundsätzen (City Planning According to Artistic Principles*, 1889) as an exemplary work of realist urbanism. Even now, we can visit numerous beautiful human-scale mediaeval urban environments introduced therein. European architects have only to return to their roots. There still exist any number of nostalgic but otherwise uninspiring tourist locales.

Japan, on the other hand, has utterly lost its own physical urban heritage. The reasons and particulars are many, but the fact is uniquely Japanese public space is extinct. In returning to physical reality, there was no going back to a nostalgic past. The important thing has been for us to see this constraint as an opportunity. We've had to come to grips with the times by making use of new sensibilities and advanced technology while relying on our own internalised body memories in order to reconstruct a physical reality. Because the path to simple re-creation of the past

にくくする技術が続々と登場したのである。自然素材に渇(かつ)えていた人々の熱意によって、建築そして広場の周辺に、大きな技術的ブレークスルーが起こったのである。

　なかでも、木材にまつわる技術には、大きな展開があった。仕上げ材料に不燃化された木材を用いることが可能になった。さらには、中規模以上の都市建築の構造材料にも、木を使う可能性が開けた。それは当然パブリックスペースのデザインの転換のきっかけを作った。繊細でやわらかな木が、町に戻ってきたのである。

　またガラスをめぐる技術にもブレークスルーがあった。ガラス自体は、20世紀の工業化によって、広く世界に広まった技術である。重たい壁がガラスに変わっただけで、人々は満足していた。しかし20世紀後半以降、プラスチックとの複合によるサイズの拡大、印刷処理による透明度のコントロールや、表情の多様化が、一気にガラスの世界を豊かにした。20世紀のいわゆる近代建築のガラスでは達成できなかったような、圧倒的な開放感、透明感、障子や布に負けないような複雑な表情がガラスの世界にもたらされたのであった。

　それがパブリックスペースを人間に取り戻すための大きな武器となったことは間違いがない。さらに、構造技術の発展も、味方となった。鉄骨の柱を木造の柱のように、細くすることが可能となった。それと木、ガラスが組み合わされた時、広場は、ぐっとわれわれに親しみやすいものへと変身した。正確にいえば、親しいものへと戻ったのである。

　このような新しい技術、素材によるヒューマン

was closed to us, the challenge for Japan has been to explore wholly new directions in public space. Japanese ideas of *hiroba* public space may even prove a salvation for a world that has reached the limits of global capitalism. Perhaps *hiroba* can serve literally as places of healing, as nurseries to give birth to social intimacy; shared casual spaces made of humble materials may even bring about the resurgence of small communities. These *hiroba* are not only beautiful comforting places, but also act as seedbeds for growing new societal structures. This book aims to show just a tiny part of these new directions.

なパブリックスペースの再生において、日本人建築家、デザイナーは世界の中で、先導的役割を果たしている。なぜなら、日本の都市空間は、本来そのような繊細な質を備えていたからである。それが、工業化、西欧化によって、一時的に失われた。しかし、日本人の中には、そのようなパブリックスペースの記憶が、しっかりと受け継がれ、流れ続けていた。そのことを、本書に収録された広場が実証している。そこには、きわめて新しい感性と技術があり、そしてなぜか、とてもなつかしいのである。

新しくなつかしい「実体」の再構築へ

　実体への回帰は、世界中の都市を覆う、大きな流れである。工業化という嵐が去り、高度成長が終わり、車自体も少しずつ都市から撤退しはじめた。この新しい状況の中で、日本が果たす役割はきわめて大きい。

　西欧では実体としての広場の実例がいくつも存在していた。かつて磯崎新は、実体論的都市論の典型としてオーストリアの建築家、カミロ・ジッテ（1843〜1903）による『広場の造型』（原著1889、日本語版1968）をあげた。そこには中世の美しいヒューマンな実体が数多く紹介されていて、今でも訪ねることができる。ヨーロッパの建築家は、ただ戻ればいいのである。なつかしくはあるけれども、その一方では退屈な観光地が並んでいる。

　しかし、日本ではいったん実体が失われた。様々な実際的な理由で、日本のパブリックスペースは一度、絶滅したのである。実体に戻ろうに

『広場の造形』カミロ・ジッテ／著

Camillo Sitte, *Der Städtebau nach seinen künstlerischen Grundsätzen*
(*City Planning According to Artistic Principles*)

も、そのままノスタルジックな昔に戻るわけには いかなかった。重要なことは、それは困難である と同時に日本人にとってのチャンスだったことで ある。時代を捕まえるあらたな感性と先端技術を 駆使して、自分の体の中にとりこまれた身体的記 憶を頼りにして、実体を再構築しなくてはならな い。過去の単なる再現という途(みち)が閉ざされていた がゆえに、日本のチャレンジは、世界のパブリッ クスペースの新しい方向を指し示しつつある。グ ローバル資本主義の限界を見てしまった世界に対 して、日本の広場は救いをもたらすかもしれな い。その広場は文字通りの癒しの空間となるか もしれないし、新しいコミュニティーを生み出す ナーサリーとなるかもしれない。やさしい素材で 作られた、さりげない広場には、小さなコミュニ ティーが戻ってくるかもしれないのである。広場 は美しく、心を安らかにするだけではなく、新し い社会構造を形成するための、貴重な苗場として も機能するのである。その一端を示すことが、本 書の目的である。

Shinjuku Southern Terrace (Shibuya-ku, Tokyo)

本書掲載の挿図の一部は、著作権法により認められた引用である。
Some of the illustrations that appear in this publication are quotes permitted under the Copyright Act.

現代日本の公共空間の実態を記録する本書の性格上、
一部写真に不特定多数の人間が写り込んでいる場合があるが、状況によって画像処理を施し、肖像権の保護に留意した。
Due to the characteristics of this publication to record the reality of public spaces in contemporary Japan, some pictures may show an unspecified number of the general public. However, the Portrait Rights were to be protected to their maximum by processing images when required.

隈 研吾 建築作品にみる「広場」

"Public Spaces" in the works of Kengo Kuma

アオーレ長岡
Nagaoka City Hall Aore

内と外、公と民が交流し合う場。
日本伝統の建築空間「ナカドマ」の概念が、
やわらかな都市の求心地を形成する

Inside and outside, a place where public and private can come together.
A concept stemming from the traditional Japanese architectural space
"Nakadoma" forms a soft urban centripetal place.

Design
隈研吾建築都市設計事務所
KENGO KUMA AND ASSOCIATES

DATA
Site: Nagaoka city, Nigata, Japan
Principal use: City hall, Assembly hall, Shops and Restaurants, Bank, Roofed plaza, Garage
Total floor area: 12073.44 ㎡
Completion: 3/2012
Structure: RC + S + Prestressed Concrete structure
Contractor: Taisei Corporation, Fukuda Corporation, Chuetsukogyo Co.,Ltd,
　　　　　　IKEDAGUMI Co.,Ltd. (Joint Venture)

18

　郊外にあった市役所を街の中心の場に復活させること。公と民の境をなくし、市民が気軽に立ち寄れる場をつくること。新潟県長岡市がめざした「まちなか市役所」の実現のため建築家が打ち出したのは、「ナカドマ＝中土間」という伝統的な日本建築の空間概念を活用するプランであった。

　土間とは、伝統的な日本家屋の主たる出入り口に設けられる、地面の土を露出したまま、あるいは「三和土（たたき）」で仕上げた屋内の場である。玄関（沓脱ぎ場）であり炊事場であり作業場であり、来客との交流の場でもあって、住居が密接した都市空間では、通行人のための路地空間にもなりうる。そんな戸外と屋内の中間、部屋のような庭のような、誰もが気軽にコミュニケーションできる場を現代に再現する試みが行われた。

　建物は、JR長岡駅とペデストリアンデッキで直結する。日本有数の豪雪地帯であるこの地に適応した全天候型の大屋根広場「ナカドマ」と周辺の回廊は、誰でも24時間利用可能でイベント会場としても人気だ。ナカドマには市庁舎などの公的機関だけでなく、カフェ等の民間施設も共存している。

　約2,250㎡のナカドマの床には昔ながらの「三和土」風の仕上げがなされ、ホールの表層には地産の杉板を用いた市松模様のルーバーが張られるなど、ヒューマンスケールに則った素材がもっぱら採用されている。

　商店街の大通りに最も近い一画にはガラス張りの議場が設置された。それはまさに"開かれた市役所"を象徴する存在である。

　かつての公共建築の代名詞といえば、大型駐車場付きの「郊外の孤立したコンクリートの箱」。ここで実現されたのはその正反対といえる、人々の生活と一体化した、活き活きとした公共空間である。「会いましょう」という意のこの地域の方言「あおうれ」から付けられた名にふさわしい、さまざまな出会いの生まれる場がこの地に根付きはじめている。

Reviving a city hall, which used to be in a suburb, by relocating it to the center of a city. Removing the borderline between public and private to create a place where citizens can come and drop by freely. Nagaoka City in Niigata wanted to have a "Machinaka (City Center) city hall", and to achieve this, the architect came up with a plan to utilize the Japanese traditional architectural space concept called "Nakadoma".

Doma (earthen floor) is an indoor place at the main entrance with a dirt floor in a traditional Japanese house. When the dirt floor has been hard packed, it is known as "Tataki[1]". Nakadoma serves as an entrance hall (where you leave your shoes), kitchen, work space, and a place for interaction with guests. It can also be an alley space for pedestrians within an urban area where houses are very close

1 Tataki: floor finishing by painting a mixture of soils, gravels, limes and bitterns, then beating and solidifying. The origin of Chinese characters used only as a phonetic symbol is mixing 3 types of materials. (Meaning of Chinese characters is three-mix-soil)

to each other. It is an intermediate space between outside and inside, which can be either a room or a garden. In this project, we intend to regenerate a place where all people feel free to communicate in the contemporary society.

The building is directly connected to Japan Railways (JR) Nagaoka Station by a pedestrian deck. Nagaoka gets one of the Japan's heaviest snowfalls and the open space with a big roof, "Nakadoma", is designed to work in all climates in this area. Anyone can use the open space and surrounding corridors 24 hours a day and this place is popular as an event venue. Public institutions such as the city hall, as well as private facilities such as cafes co-exist in front of Nakadoma.

On the floor of Nakadoma, which is approximately 2,250㎡, a traditional "Tataki" style finishing was applied. Materials in accordance with human scale were mainly adopted: for example, the outer layer of the hall has checkered louvers made of local cedar wood boards.

A glazed assembly hall was placed at the corner closest to the main street of a shopping district. It is a symbol of an "open city hall".

One might describe old public architecture in Japan as "an isolated concrete box in a suburb" with a large parking space. What we realized here, Nagaoka City Hall Aore, is the opposite type of public architecture which has a lively public space that integrates with people's lives. "Aore", derived from "let's meet" in a local dialect, is an appropriate name for this new place of various encounters which has started to take root in the area.

la kagu

訪れる者も通り過ぎる者も
歩みの速度が自然とゆるむ、
木漏れ日の落ちる大階段

Large stairs with dappled sunlight filtering through the leaves of trees
invites both visitors and passers-by to slow down and relax
at their own pace.

Design supervision
隈研吾建築都市設計事務所
KENGO KUMA AND ASSOCIATES

DATA
Design: SHIMIZU CORPORATION
Site: Shinjuku-ku, Tokyo, Japan
Principal use: Store building, event venue
Total floor area: 962.45 ㎡
Completion: 10/2014
Structure: S structure
Contractor: SHIMIZU CORPORATION

　古い倉庫を街といかにつなげるか。それが最大の課題であったという。
　歴史と文化の息づく街・東京は神楽坂駅前に位置する築40年超の倉庫を、ファッション・生活雑貨・カフェ・家具・本などの複合商業施設として生まれ変わらせる計画である。
　施主である新潮社の「本の倉庫」であった建物は、汚れてはいたが波板スレート張りの外観、あやとりのようなトラス構造の天井組など、50年近くを経過した建物ならではのユニークな特徴を備えていた。それを最大限活用するために、傾斜面に立つ現状の地形に沿わせるトポロジカルなウッドデッキを取り付け、建物の入り口側には大開口が設けられた。これにより、駅からの「歩道

The biggest challenge was how to connect an old warehouse and a town.

The plan was to revive a warehouse, over 40 years old, located in front of Kagurazaka Station in Tokyo, where history and culture are brimming, into a complex commercial facility including fashion shops, homewares shops, cafes, furniture shops and book stores.

The building used to be the "book warehouse" of the client, Shinchosha publishing company. It was dirty, but had some characteristics unique to a building with a history of nearly 50 years, such as an external wavy slate and ceiling frames with truss structure like cat's cradle[1]. To maximize these characteristics, a topological wooden deck along the current sloping geographical feature was installed and a large opening at the entrance side of the building was created. Per this plan, a pathway to the building was smoothly connected by a deck in the same way as an "extension of pathway" from the station and visitors are led to the building seamlessly. We can enter and exit the lecture space which has a capacity of just under 100 people located in the 2nd floor directly via the deck stairs.

The wooden deck stairs are made of Selangan batu, natural wood with high durability. Widths of stairs seem random at first, but they are actually designed with much precision and intention.

In good weather, people find their favorite spot, sit and take a break, using the stairs as benches continuously.

Three existing trees, including a fairly large cherry blossom and paulownia trees were utilized. Dappled sunlight, filtering through the leaves of these trees, falls on the deck plays a role in providing a comfortable space.

The deck of la kagu is an urban public space which can be called a "large bench". la kagu is a commercial space attracting not only people with a direct purpose of shopping but also people without any particular purpose. It invites people without any purpose to stay and allows them to "just be there". It is an unexpected and rare occurrence for modern cities.

1. Configurated structure type by utilizing stability of triangle

「の延長」のように、デッキによって建物への道はゆるやかにつながれ、来訪者はスムーズに建物に導かれる。ウッドデッキの階段はそのまま2階へと続き、100名弱を収容できるレクチャースペースに直接出入りすることも可能になっている。

　天然木で耐久性の高いセランガンバツ材を用いたデッキの階段は、場に応じてその幅が変えられている。一見ランダムに見えるその裏にも、緻密な計算・意図が込められている。天気の良い日には、自身のお気に入りの場を見つけ、ベンチとして階段に腰掛け、休息をとる人が後を絶たない。

　もともとこの地に生えていた3本の樹木、大ぶりの桜の木と桐の木も伐採の憂き目を見ずに活用された。デッキに落ちる木漏れ日が、心地よい居場所の提供に一役買っている。

　la kagu のデッキは、それ自体が「大きなベンチ」とも呼べる都市の広場空間である。ショッピングという直接の目的を持つ人だけでなく、目的のない人も呼び込み、留まらせることができる、「ただなんとなく居る」ことが許容される商業空間。それは思いのほか、現代都市にとって希少な存在である。

現在の『GINZA KABUKIZA』は、1889年創設の初代歌舞伎座から数えて5代目となる。外観や客席など、吉田五十八の設計による第4期歌舞伎座を生かしつつ、岡田信一郎設計の第3期から続く桃山様式が踏襲された。

計画は単なる復元にとどまらず、地下2階には東京メトロ東銀座駅と連続する広場が新設され、5階には屋上広場（28・29頁）も整備。いずれも劇場チケットがなくても自由に出入りでき、都市との新たな連続性が生み出されている。

新設された地下広場の名は「木挽町広場」。かつてこの地一帯は、江戸城普請に従事した木挽き

The current "GINZA KABUKIZA" is the 5th KABUKIZA since 1889. It made the best use of the exterior, seats and the like of the 4th KABUKIZA designed by Isoya Yoshida while following the Momoyama Style used in the 3rd KABUKIZA designed by Shinichiro Okada .

The plan of the 5th KABUKIZA was not only to restore the building. A new public space connecting to Tokyo Metro Higashi Ginza Station was established in the 2nd basement floor and the 5th floor was also equipped with a rooftop public space (pp. 28- 29). To enter these two spaces, theatre tickets are not required and thus a new continuity

Design
隈研吾建築都市設計事務所
KENGO KUMA AND ASSOCIATES

DATA
Site: Chuo-ku, Tokyo, Japan
Principal use: Underground shop for the theater, Resting place, Subway passageway
Total floor area: 12073.44 ㎡
Completion: 4/2013
Collaborative design : Mitsubishi Jisho Sekkei Inc.
Contractor: SHIMIZU CORPORATION

木挽町広場（歌舞伎座）
GINZA KABUKIZA, Kobikicho Plaza

都市の延長としての劇場、
劇場の延長としての都市。
両者をつなぐ、緩衝地帯のような広場。

Theater as an extension of a city
City as an extension of a theater
Public space acting as a buffer zone connecting both

職人（製材業者）が多く住むことから木挽町と呼ばれ、また芝居茶屋が蝟集する町だったことから「木挽町に行く」といえばそのまま「芝居見物に行く」ことと同義語であったという。今回新しい広場に付けられた「木挽町」の名には、かつての賑わい溢れる江戸の芝居町を現代に、という期待が込められているのである。

広場は客席の真下に位置するが、地下から劇場には直結していない。劇場に入るには、いちど地上に出て正面玄関に回る必要がある。それは、歌舞伎を見るという「ハレ」の行いを損なわせないための演出でもあるのだ。地下から地上に出て、雨でも濡れない深い庇の通路を抜け、唐破風をくぐる。一連の動線を進むことは、まるで日常のケガレを落とす禊ぎの儀式のようでもある。地下鉄空間から広場に入る際も、灰色から鮮やかな赤へと、色彩に明確な変化がつけられ、「ハレの場」に来たことを意識させられる。

壁面上部を総鏡張りにした開放感溢れる空間のなか、仲見世のように立ち並ぶ土産物店に幕間の客がつめかける。広場中央に置かれた提灯のシンボル性も高く、観劇客はもちろん、通行人も思わず足を止める、魅力溢れる都市空間が形成されている。

with city was created.

The new 2nd basement public space is called "Kobikicho Plaza". This area was called "Kobikicho" before because a lot of Kobiki Shokunin (lumber producers) lived there, who engaged in the construction of the Edo Castle. Theater tea houses were also in abundance in the town, which lead to the phrase "going to Kobikicho" meaning "going to the theater". By naming this new public space "Kobiki" it creates the expectation that a lively Edo theater town would be revived in modern times.

Although the public space is located directly below the theater seats, it is not directly connected to the theater. If we want to go to the theater, we have to go up to the ground level and then go around to the front entrance. This was done so as not to detract from the important experience of seeing Kabuki. As we go up to the ground level from the basement, we go through a passage below deep eaves, providing protection from rain, and go under the Karahafu (a style of gable). Passing through this chain of spaces is like having a Misogi (purification ceremony) to wash out daily uncleanliness. We can also sense that we are coming to an important place when we go into the public space from the subway space because the color of the spaces change clearly from grey to vivid red.

One space has an upper part of the wall fully mirrored to produce a sense of openness, where the audience rushes to souvenir shops, standing a line like the Nakamise street district during intermission. A lantern placed in the center of the public space is highly symbolic and an attractive urban space is formed, where not only audiences but also pedestrians stop on impulse.

マルセイユ現代美術センター
FRAC Marseille

ストリートの自由さを持つ
浮遊するパブリックスペース

*A floating public space
With the flexibility of a street*

Design
隈研吾建築都市設計事務所
KENGO KUMA AND ASSOCIATES

DATA
Site: Marseille, France
Principal use: Museum, Conference room, Housing, Document center, Office, Cafe
Total floor area: 3895㎡
Completion: 12/2012
Structure: S + RC structure

FRAC (FOND Régional D'Art Contemporain) は次世代芸術家の育成と新しいアートの創造を目的とする地域密着型の組織である。その理念に従い、従来の閉じた箱としての美術館に代わる、地域に開かれた文化施設が計画された。ストリートがそのまま展示空間やワークショップの空間になるような設計イメージである。

この立体化されたストリートの結節点として、ビル中層部に大きな開口部を持つ空中テラスが設けられた。そこはアート製作・展示・パーティにも使われる多目的空間である。このテラスが「大きく・重い」建築物の既成概念を覆していることは一目瞭然だが、外壁には約1,500枚のガラスパネルが用いられ、千鳥状の構成が、遠目に見ると紙が舞っているような、さらなる軽やかさを建築に与えている。

重厚な建築の対極に位置する、室内・室外が連動した、都市に開かれた浮遊する美術館。地域にアートを伝えるパブリックスペースを、という理念の実体化である。

FRAC (FOND Régional D'Art Contemporain) is a community-based organization with the purposes of cultivating the next generation of artists and creating new art works. Following this principle, a community cultural facility was planned instead of the conventional closed box museum. The building is designed so that the street itself becomes an exhibition space or a workshop space.

A hanging terrace with a large opening was created in the middle of the building as a node point for this three-dimensional street. The terrace is a multipurpose space for art production, exhibitions and events. It is obvious that this terrace overturns the stereotype of "big and heavy" architecture. Approximately 1500 pieces of glass panels are used and the zigzag structure gives the impression of lightness; the structure looks like floating paper from a distance.

FRAC Marseille is a floating museum opened for the city, which is located opposite to profound architectural structures, and it connects the inside with outside. It substantiates the notion of a public space expressing arts to the community.

Design
隈研吾建築都市設計事務所
KENGO KUMA AND ASSOCIATES

DATA
Site: Besançon, France
Principal use: Art center
Total floor area: 11,925㎡
Completion: 12/2012
Structure: S + Timber structure

ブザンソン芸術文化センター
Besançon Art Center and Cité de la Musique

人間と自然を媒介する場
A medium to connect humans and nature

　フランス東部の都市ブザンソン。中心部を流れるドゥ川の河岸の敷地に、美術館、音楽学校等の複合的文化施設を作る計画である。

　美しい河岸に複数の箱状の建築物が建てられ、その箱の間に木製の大きな屋根がかけられた。屋根には植栽、地元の木、石、ガラスなどがモザイク状に配置され、そのモザイクの落とす影が木陰の多い公園のような空間を生み出している。川とゆるやかに連結されたその緩衝地帯（＝屋根の下の孔）を通し、人々は心地よい風と木漏れ日という自然の恩恵を存分に享受できる。

　近くの観光地である古い城砦の上からの眺めも、この第5のファサード＝屋根のデザインを決めた一因であるという。それぞれの箱の外壁にも木板が張られ、ここでも板と板の隙間が、建築に軽やかさと開放感を与えている。

　建築は孤立したオブジェであってはならない、自然と人間を「媒介」するものであるという設計者側の哲学が、この隙間（孔）の一つひとつに投影されているのである。

Besançon City in the east France — The plan was to construct a complex cultural facility including a museum and music school at the grounds of the river Doub which passes through the city centre.

　Multiple architectural boxes were constructed on the beautiful river beach and large wooden roofs were put on top of these boxes. Plants, local trees, stones, glass and the like were placed on the roofs in a mosaic style and shadows produced by this mosaic make the space feel like a park with dappled sunlight in the shade of trees. People can sense nature's blessing as a pleasant wind blows softly through the sunlight filtering through the leaves of trees in this buffer zone (=void under the roof) connecting loosely to the river.

　A view from the top of an old fortress, which is a nearby sightseeing spot, was a factor in addressing the roof design as the 5th facade. The outside walls of each box are finished with wooden boards and gaps between boards give lightness and a sense of openness to the structure.

　Architecture should not be an isolated object, but it should serve as a catalyst connecting nature and people. This philosophy of the architect is projected onto every single gap.

はじめに ―日本らしい広場をめざして―

　都市の広場に多様な人々が集い、いろいろな活動が繰り広げられる光景は、見ていても楽しい。また、都市を経済的にも文化的にも元気づけるのに、そうした広場の果たす役割は大きい。特に、ヨーロッパを旅すると、今も活気に満ちている素敵な広場との出会いをしばしば体験できる。

　とはいえ、都市空間の中における人間の振る舞いかたというのは、国民性によってさまざまに異なる。広場に最もふさわしいのは、ヨーロッパの中でも、やはりイタリア人だろう。まわりを華麗な建築で囲われた堂々たる広場で、大きな身振り、手振りでコミュニケーションし合う彼らを見ていると、まるで演劇そのものといった感じがする。

　私が最も好きな広場の一つが、ヴェネツィアの本土側にある小都市、トレヴィーゾの**シニョーリ広場**（図1）だ。水路が巡る魅力溢れる街なのに観光客が少なく、この広場でもまさに市民が主役なのだ。日中は、幼児を連れた若い母親たちが仲間とお喋りを楽しむ一方、リタイアした年配の男たちがカフェでのんびり時を過ごすのだが、昼食前と晩食前の時間帯ともなると雰囲気が一変する。大勢の市民がここに集まり、仲間との立ち話の輪が無数にでき、広場全体が屋外の壮大な社交場と化す。

　こういったイタリアの広場は一種の憧れの場所であっても、そこに我々が立つと、何かちょっと気恥ずかしく、落ち着かない感じがするというのが正直なところだ。実際、「日本には広場が発達しなかった」と長らく言われ続けてきた。政治・社会の仕組みにその原因があるのはいうまでもな

Introduction
—Working Toward Unique Japanese Public Spaces—

It is joyous to watch various people gather in urban public spaces as a variety of activities unfolds. These kinds of public spaces also play a large role in supporting the economic and cultural activities within cities. This can be witnessed especially when travelling in Europe where one encounters wonderful public spaces full of energy.

　Indeed, the way that people interact with public spaces is dependent on national characteristics and in my opinion, those whose nationality is most suited for public spaces within Europe are the Italians. They communicate with each other using grand body and hand gestures in brilliant piazzas surrounded by splendid buildings to create scenes reminiscent of a play.

　One of my favorite piazzas is the Piazza dei Signori (Fig. 1) in the small city of Treviso, located on the mainland of Venice. Treviso is an extremely attractive city with waterways, not crowded with tourists, and here you can see the citizens play the leading characters in this public space. During the day time, young mothers with toddlers enjoy chatting with friends while retired elderly men relax at cafes. However, the atmosphere of the piazza changes dramatically just before lunch time and supper time. Throngs of people come to the piazza forming countless circles of friends standing and talking. The whole public space becomes a magnificent outside social meeting place.

日本独自の広場、その多様性の検証

陣内秀信 （法政大学デザイン工学部教授・東京大学工学博士）

Diversity of Unique Japanese Public Spaces

Hidenobu Jinnai (Professor of Faculty of Engineering and Design in Hosei University, PhD in Engineering)

いが、それに加えてこうした人々のメンタリティの違いもあったと思われる。

それに対し、ヨーロッパでは古代ギリシアの都市に広場がまずは発達し、中世には現在に直接つながる広場の形式が確立した。こうした広場の存在は、市民の自治の象徴という意味も伴って、日本の人々にとっての一つの願望の対象となった。だが、我々の近代に、ヨーロッパの都市の広場にインスピレーションを得て、それを応用・導入しようとしても、そう簡単にいかなかったことは、歴史が物語っている。

といって諦める必要は毛頭ない。本稿では、日本の都市にも人々が集まり交流・交換する場所が実はいくつもあり、日本版の広場が多様に存在してきたことを検証してみたい。そのためにも、日本の都市の空間を、日本らしい文化風土、人間関係、空間人類学の視点からも考察する必要がある。しかも、嬉しいことに、豊かな時代を迎えた1980年代後半から、都市開発の手法も大きく変わり、いかにも現代日本らしい創造性に富んだ価値ある広場が続々と誕生してきている。その状況にも目を向け、最近できた日本の広場の特徴とその意義について考えてみたい。

ヨーロッパの広場史概観

とはいえ、広場論を展開するには、その前提として、歴史的経験の長いヨーロッパの広場を知る必要もある。まずは、市民による民主主義が生まれたギリシアの都市に、アゴラという広場が誕生した。広場が大好きだった建築家・丹下健三の授業を大学で受講したことがあるが、その第1回目に、ギリシアのアゴラが登場したのをよく覚えている。彼の原点なのである。

Although such Italian piazzas are wondrous and dream-like for Japanese people, the truth is, we feel a little bit awkward and restless when we are actually there. In fact, we have often said that public spaces were not developed in Japan until recently. Japanese political and social systems are the obvious reasons for this, but another reason could also be the different mentality of the Japanese people.

Plazas were first developed in the ancient Greek cities of Europe. The medieval period saw the establishment of public spaces which directly relate to the current system of European public spaces. Such spaces, seen as a symbol of self-government, become desirable amongst Japanese people. However, as history would show, the inspiration created by European public spaces would not be enough to introduce them into Japan.

However, we should not dismiss the idea of Japanese public spaces. In this article, I would contend that actually we have a lot of urban places where people gather and share exchanges in Japan and, indeed, Japanese style public spaces have existed in various forms. To that end, I need to examine urban spaces in Japan by considering the unique Japanese cultural climate, human relations and spatial anthropology. Moreover, the good news is that since the wealthy times of the late 1980's, urban development techniques in Japan have changed dramatically and worthy public spaces which support rich contemporary Japanese creativity have been created one after another. After reviewing an outline of European public space history, I will turn your attention to the characteristics and meanings of contemporary Japanese public spaces.

Outline of European Public Space History

As a starting point in discussing public spaces, we must have some knowledge about the long history of the European experience with public spaces. First, a public space called Agora was established within the ancient cities of

図1
シニョーリ広場（イタリア）
Figure 1
"Piazza dei Signori" (Italy)

アゴラの柱廊で囲われた形式は、ローマ時代のフォルムに受け継がれ、都市の公共空間を形づくった。そこには市民の住宅はなく、公的な政治・行政の機能、神殿・宗教施設に加え、市場・店舗の経済機能、演劇・文化の活動の場でもあった。だが、ローマ帝国の崩壊とともに都市が衰退し、その後、中世における都市の復興の時代にこそ、現在に繋がる広場の原型ができる。古代にすでに都市があれば、その中心広場、フォルムを受け継ぎつつ、市庁舎の聳える象徴的な広場が形成された。そのまわりには、店舗や職人の工房ばかりか、市民の住まいも取り巻いている。市場もこの市庁舎広場に寄り添う。今日につながる市民意識、自治の精神は、この中世都市の広場と結びついて生まれたといえる。

ルネサンス時代、芸術文化のよき理解者・君主のもとで、建築家も関与して広場が設計され、より規則的、左右対称、幾何学的な広場を実現した。ルネサンスの広場は、宮廷文化を反映し、壮麗な祝祭の舞台ともなったが、市民・住民の自治の舞台という性格はむしろ弱まった。ルネサンスの広場は、人間の身体寸法による閉じた空間を尊重したが、続くバロック都市では、より大きなスケールで都市の街路と繋がり、動きのあるダイナミックな造型が実現した。

だが、その西洋でも、近代になると車社会を迎え、人間のための広場が解体しはじめた。19世紀末、ウィーンで活躍したカミロ・ジッテ（13頁）が文明批評的にそれを批判する一方、中世の広場を評価し、大きな影響を与えた。

結局やはり、中心に華やかな都市中心としての広場があるのは、世界でも西洋にしかない。それも、建築と同様、ギリシア・ローマ・中世・ルネサンス・バロックと造形の様式で分類できるのだ

Greece where the democracy was founded by its citizens. I have taken university classes lectured by Kenzo Tange who was a big fan of public spaces and I clearly remember that he was discussing Agora in his first class. The agora was the origin of his architecture.

Agora was surrounded by colonnades. That style was inherited by the Imperial forums of the Roman period which formed urban public spaces. These fields were public politics and government administration functioned. They also serviced the economy with markets and shops and provided a sanctuary where religious, theatrical and cultural activities took place. Residences, however, were not located at these fields. The prototype of the public space which is truly related to current public spaces was created during the era of urban reconstruction in medieval times after had been in decline at the fall of the Roman Empire. Symbolic plazas with city halls were created while inheriting those Imperial forums which formed the concourses of cities in ancient times. Around these public spaces were not only shops and artisan studios but also residences. Markets were also found close to the public spaces of the city halls. We can say that the current public awareness and spirit of self-government was established with reference to the medieval urban public space.

In the Renaissance period, more systematic, bisymmetric and geometric public spaces were designed with the participation of architects under supporters of art and culture and sovereigns. The public space of the Renaissance period reflected its court culture and became a stage for magnificent celebrations; however, this coincided with a decline in the use of these spaces by ordinary citizens for their own self-governing. In cities following Baroque period, public spaces were connected with urban streets on a bigger scale and were shaped more dynamically by motion.

Nevertheless, even in this Western culture public spaces for people started to be dismantled in the modern times. At the end of the 19th century, Camillo Sitte (See p. 13), who played in an important role in Vienna, criticized this change on civilization. On the other hand, it was his evaluation of medieval public spaces which created a big impact.

The public space as a bright urban center in the middle of a city existed only in Western countries. Additionally, the culture of public space is sometimes seen as a unique privilege of Western cities, reflecting Greek, Roman, medieval Europe, Renaissance and Baroque architectural styles. In fact, we seldom, if ever, see magnificent public spaces which could be classified by the art and culture styles of these periods in Japan. For this reason, an entirely different

から、広場の文化は西洋都市の特権のようにも見える。実際、こうした時代の芸術様式で分類されるような立派な広場は日本にはほとんど見当たらない。というわけで、日本の広場をダイナミックに面白く語るには、別の枠組みでアプローチすることが必要となる。

日本的な広場の系譜を探る試み

　日本に広場がなかったかというと、決してそんなことはない。人が集まる公的な性格をもった場所、都市空間という風に考えを広げてみると、とたんに、たくさんの面白い候補が思い浮かんでくる。神社や寺院の境内や門前に、橋のたもとや河原の水辺に、そして市場に、そうした空間が多様に成立していた。中世を専門とする歴史家の網野善彦（1928～2004）はかつて、こうした場所に権力の手が及ばない、守られた「アジール」（自由な場）が成立し、そこに我が国の都市の萌芽が生まれたとした。広場のあり方は、このように都市の歴史そのもののありかたと深く結びついているのである。

　西洋の発想から自由になって、日本独自の広場像を描く試みに最初に取り組んだのが、伊藤ていじ（1922～2010）が中心になって編まれた『日本の広場』（『建築文化』1971年8月号の特集を書籍化したもの。都市デザイン研究体・彰国社・2009年）で、今、読んでみても新鮮な内容である。イタリアに留学し、都市の広場を研究した加藤晃規（関西学院大学教授）は、日本に戻ってからは、比較の視点に立って、日本の都市空間の中に広場的なものを見つけ出し、「場所的広場」と命名して、そのありかたと特徴を興味深く考察した。

framework is required to meaningfully discuss the topic of public spaces in Japan.

Exploring the Linkage of Japanese Public Spaces

In actual fact, public spaces are to be found in Japan. When we start to expand the idea of the public space as a place or urban space, with public characteristics where people congregate, a lot of interesting places come to mind. Such spaces were created on the premises of shrines and temples, in front of their gates, at the foot of a bridge, next to rivers or at a marketplace. Yoshihiko Amino (1928-2004), who is a historian specializing in medieval Japanese history, has said that these places provided "asylum" (safe havens), out of the reach of authorities; the seeds from which Japanese cities sprouted. In this way, the form of public spaces directly relates to the prevailing urban culture

"Public Spaces in Japan" (originally from a special article in the August 1971 issue of "Kenchiku Bunka" and subsequently published as the book, Urban Design Movement, Shokokusha, 2009), which was complied mainly by Teiji Ito, was the first book which attempted to describe a picture of the unique Japanese public space, as distinct from the Western one. Its content still feels new to me. Akinori Kato ((Professor of Kwansei Gakuin University) went to Italy to study urban public spaces. After returning to Japan, he compared countries and he was able to find places in urban Japan functioning as public spaces. He labeled these "places as public spaces" and provided an interesting examination of their state and characteristics.

In "Public Space" (Space Design series) published by Shinnippon-Hoki Publishing, I also broke down the framework to understanding Western style public spaces, classifying them by function and role, rather than form, shape and style. Using this approach, I was able to show that we do indeed have places as public spaces in Japan. These can be found by rivers, at temples and shrines, at crossroads and elsewhere. They emerged during the Edo period (early modern times) and I introduced the idea that they have continued as public spaces into modern times and we can even attach different meanings and aspects to these spaces in Japanese cities everywhere.

Sacred Places as Public Spaces

In fact, we have a lot of places with a scaled to match the Japanese sense of the body: places for cultural exchange established to meet social needs and gathering places for peo-

私自身も、新日本法規刊行の『広場』（SDシリーズ）において、西洋流の広場理解の枠組みを壊し、形態・造型・様式から捉えるのではなく、機能・役割から見た広場の類型化に取り組み、日本にも広場的な価値ある場所がたくさんあることを論じた。橋のたもと・門前・境内・河原・四つ角（辻）など、すでに江戸時代（近世）から登場し、成立してきたものが、近代になっても骨太に持続し、新たな意味や様相をも加えて、今なお日本の都市のあちこちに見られることを紹介した。

聖なる場所が広場に

　実際、日本には我々の身体感覚に合ったスケール感をもち、社会的なニーズと対応して生まれた交流の場、人々にとっての集合的な場がたくさんある。神社の境内も伝統的な広場の一つである。

東京のような現代の大都会でも、地域の神社の境内の一角にベンチが置かれ、老人たちが居心地よさそうにくつろぎ、のんびり会話を楽しんでいる光景をよく目にする。周りの樹木も落ち着きを与える。児童遊具が置かれ、子供たちがそこで飛び回っているということも、よくある。日本の典型的な広場のシーンの一つだ。かつて我々の世代は、小学校が終わると、いつも神社の境内でゴムボールの野球に精を出したものだ。

　これらがどちらかというと静的な広場ならば、動的な広場の代表は、縁日の日の**とげ抜き地蔵**（124頁）の周辺に展開するような空間だろう。前述の『日本の広場』にも登場する、巣鴨の小さな宗教空間である高岩寺には、普段は何ということもない境内の姿が見られる。ところが、4のつく縁日にあたる日となると、様相が一変する。門前の商店街いっぱいに露店が並び、原宿の竹下通り

ple. Premises of shrines are one of these traditional sacred public spaces. Even in a modern metropolis, such as Tokyo, we often see the scene of old people relaxing and enjoying conversations on a bench near the corner of a shrine, surrounded by trees and peacefulness. We can also see the scene of children playing around play equipment areas. These are the typical scenes to be found at public spaces in Japan. When my generation was young, we would play baseball with a rubber ball on the premises of a shrine after elementary school.

　While these might be considered rather static public spaces, we can also find dynamic public spaces in Japan, an example of which is the space around Koganji Temple, also known as Togenuki Jizo (p. 124), on the day of a fair. Koganji Temple is a small religious space in Sugamo, Tokyo, giving the usual appearance of a Japanese temple. However, we witness a complete change on the day of a fair, held on special dates (including those days of the month containing a 4 (4th, 14th, and 24th)). The shopping district in front of the temple gate fills with street stalls and is descended on crowds of shoppers. It is sometimes noted that this crowd is similar to that of the famous Takeshita-dori Street in Harajuku, except for one important difference; most of the shoppers here at Koganji temple are old

ladies. Thus, the very apt naming of Koganji temple as the "Harajuku of old ladies", because they seem radiant with joy, just like girls in Harajuku. The premise of the temple becomes "standing room only". It is a space for stalls, markets, various performances and activities such as prayers and fortune telling which serves as an ideal prototype for public space in Japan.

　The premise of Hanazono Shrine in Shinjuku is also a multipurpose public space; a sacred place that sometimes changes into a lively place of celebration (p. 122). On days (Ichi no tori, Ni no tori and San no tori which are the first, second and third days of the rooster) of the Tori no ichi (rooster market) in November every year, countless lanterns forms a line and street stalls selling rakes for "good luck and thriving business" also line up. A well known play at the "Aka-tento" (red tent) by the Karagumi Theater Company, which dashed out from a theater in the 1960's and is lead by Juro Kara, is often performed here too.

　Takigi-Noh (Torchlight Noh) (p. 116), which is performed on the premises of a temple as a static public space, also produces special effects because it is in a quiet and dense forest. I have experienced torchlight Noh at both the Hie Shrine in Akasaka and the Meiji Shrine. By presenting Noh (a traditional Japanese musical drama) in this special

さながらの雑踏で、もみくちゃにされる。ただこ こに集まるのは年配の女性ばかりだ。「おばあち ゃんの原宿」というネーミングは実に当を得てい た。原宿に集まる女の子たち同様に、おばあちゃ んたちの顔は輝いている。境内は足の踏み場もな いほどにぎっしりで、願掛けをはじめ、祈禱、占 いなどさまざまなパフォーマンスの空間となる。 あらゆる商品を売る市も立つ。日本の広場の原型 を見る思いだ。

　新宿の花園神社の境内もまた、聖なる場がしば しば賑やかな祝祭の場に転ずる多目的の広場とい える（122頁）。毎年11月、酉の市（一の酉、二 の酉、三の酉）の日には、境内に無数の提灯が並 び、開運招福、商売繁盛の熊手を商う露天商が境 内にずらりと並ぶほか、花園神社名物の見世物小 屋の興行が行われる。さらには、1960年代に劇 場から飛び出し、演劇の世界を変えた唐十郎が率

いる劇団唐組の紅テント芝居が今でも、よくこの 境内で催されている。

　静的な広場である寺社の境内で行われる薪能(たきぎのう) （116頁）もまた、その静寂、深淵なる森が故の、 特別な効果を生む。私も赤坂の日枝(ひえ)神社、明治神 宮で薪能を体験したことがあるが、夕暮れから闇 に転ずる時間、坂や階段を上り、あるいは長い砂 利の参道をしめやかに進み、異界に入って、かが り火で照らし出される舞台に日をやるという空間 の演出は、聖なる広場の出現を思わせる。

人の流れが生む、辻と橋のたもとの広場

　建築のモニュメンタリティよりも、場における 人々の活動、営みに大きな意味がある日本の都市 では、昔から今に至るまで、人の流れが重要であ る。ヨーロッパの都市のように、道があらゆる方

---

type of public space, torchlight Noh gives us the impression of emerging from a sacred place. As we move from dusk into darkness, it feels as though we're going up a slope or a flight of stairs, or we're quietly approaching the spirit world along a gravel road. Then we turn our eyes to the stage lit up by a sentry fire, which gives us an impression that a sacred space has emerged.

Public Spaces in Crossroads and at Feet of Bridges Produced by a Flow of People

The flow of people is important in Japanese cities to this day, where the activities and actions of people have a greater meaning than the monumentality of architecture. It is different from European cities where streets converge to one point from all directions to a central public space surrounded by majestic architecture. Japan tends to have small public spaces at a flow node of people who are moving within urban space and forming a lively atmosphere.

　This is how the importance of "crossroads" space was established. A statue of Jizo (the guardian deity of children) or a Hokora (a small shrine) are often placed at these crossroads. Even in the modern times, intersections are especially important places in Japanese cities and a lot of popular public spaces are located at intersections. In Tokyo, consider Ginza. The 4-chome corner, where Wako is (p. 130), has been a symbolic spot of the town since the early Meiji period when Ginza Brick Street was created. The clock tower of K.Hattori was also constructed here and the corner became popular as a landmark of Japan's Western movement. The corner is often used as a meeting to this day.

　Sony Building at Sukiyabashi Crossroad (Fig.2) is also a masterpiece of street corner architecture. At one corner of this building, they keep a small but interesting public space for seasonal events. It is a superb result of an urban "trick". Street corners which change aspects depending on seasons truly suit the Japanese mentality.

　It would seem that medium sized public spaces with pocket park characteristics, which have moderate tricks and include variability and stimuli, suit Japanese people more than the Western European style large public spaces. This explains why small public spaces in front of Laforet Harajuku (Fig.3) or Kinokuniya (Fig.4) in Shinjuku have been popular.

　As per the significance and meaning of crossroads mentioned above, "streets" themselves also play the role of public spaces in Japanese cities. Before the arrival of cars, street spaces provided lively scenes, filled with street signs

向から一点に集まってきて、そこに堂々たる建築が囲む求心性の高い広場が成立するというのとは事情が違う。人が都市空間を移動する流れの結節点に、ちょっとした広場が形成され、賑わいの場が生まれるという傾向が見られるのだ。

「辻」の空間の重要性もこうして生まれた。そこに地蔵や祠が置かれることも多い。現代でも、交差点は都市の中で決定的に重要なスポットであり、人気のある広場のかなりのものが、そこに成立している。東京なら、銀座を思い起こすとよい。4丁目の和光の角（130頁）は、銀座煉瓦街がつくられた明治の初期から、街の象徴的なスポットとなり、服部時計店の時計塔がつくられて、文明開化のランドマークとして人気を集めた。今も、待ち合わせ場所としてよく使われる。

数寄屋橋の**ソニービル**（図2）も街角建築の傑作だ。コーナーにちょっとした空地を残し、小粒ながら面白い広場を設けて、そこで四季折々のイベントを行う。まさに都市の仕掛けとしては最高の出来映えだ。季節に応じて表情を変える街角は、日本人のメンタリティに実に合っている。

どうも我々には、堂々たる建築で整然と囲われた大規模でガランとした西欧的な広場よりも、ほどほどの大きさで、仕掛けが適度にあり、変化や刺激をも含んだようなポケットパーク的な広場が身に合っているようだ。原宿のラフォーレ前（図3）や新宿の**紀伊國屋書店**（図4）の前の小さな広場が人気を集めてきたのも、よくわかる。

辻の意味をすでに述べたが、そもそも日本の都市では、「道」そのものが広場の役割をもっていた。車のない時代、町家の店の前には床几（しょうぎ）が出され、路上に商品が溢れ、看板も置かれ、道空間は賑わいをもった。町家には人々が住みながら商売をしたから、生活感が道に溢れ、子供も路上で遊

and stalls offering an array of products for sale outside of shops. People also lived at the shops and children played in the streets. An alley without cars became the best place for children to play. The alley as a public space provided a variety of uses for children; a space to play "kick the can", 3-base baseball, marbles or to share picture-story shows. At the back side of Shinagawa-shuku, there is an area called Koizumi Nagaya, elaborately filled with skirting alleyways. We can still find some wells in these wide alleys and the wells can tell us how deep the community connection was.

In particular, space at the foot of a bridge became as important as those crossroads on land. This was a node point of transportation. In fact, the largest public space in the Edo period arose at a broad street at the west foot of the Ryogoku Bridge crossing the Sumida River (Fig.5). A flow of people both on the water and on the roads gathered here and this public space had aspects of a party area full of stimuli. Temporary tea houses and theater lined up and street performers and peddlers milled around the empty space which was reserved for evacuation from fire and as a firebreak: the space formed a Japanese style public space with chaotic energy.

In Edo towns, we could see similar these scenes at empty spaces located at the foot of the Edo Bridge and the public space at the Sujikaigomon Gate (Fig. 6) at the Kanda River.

Even in the modern times, the importance of a foot of a bridge was inherited. In the early Taisho period, fully-fledged Western style buildings including Murai Bank and

図2
ソニービル前（東京都中央区）
Figure 2
Sony Building (Chuo-ku, Tokyo)

図3
ラフォーレ原宿前（東京都渋谷区）
Figure 3
Laforêt Harajuku
(Shibuya-ku, Tokyo)

図4
紀伊國屋書店新宿本店前（東京都新宿区）
Figure 4
Kinokuniya (Shinjuku-ku, Tokyo)

図5
歌川広重『両国橋大川ばた』（名所江戸百景より）
Figure 5
Hiroshige Utagawa :
"Ryogoku-bashi Okawabata" from
"One Hundred Famous Views of Edo".

んだ。車の入らない路地ともなれば、一層、子供の遊び場として格好の広場となった。缶蹴り、三角ベースの野球、ビー玉、そして紙芝居など、路地広場の使い道は実に多様だった。品川宿の裏手に、小泉長屋と呼ばれる路地が複雑に巡るエリアがあり、その広めの路地の所々に、井戸が残っていて、コミュニティの繋がりの深さを物語る。

交通の結節点という意味で、陸側の辻と同じように、水辺の橋のたもとの空間が、とりわけ重要な役割をもつことになった。かつて江戸最大の広場は、実は、隅田川に架かる**両国橋**の西のたもとの広小路に生まれていた（図5）。水上と道路の両方での人の流れが、ここに集まり、刺激に溢れる盛り場の様相を示していた。火除け地としてとられた空地に、仮設の茶屋や芝居小屋が並び、辻芸人、物売りがひしめいて、猥雑な活気に満ちた日本的な広場を形づくっていたのだ。

江戸の町では、江戸橋広小路にも、神田川の**筋違御門の広場**（図6）にも、こうした賑わいが生まれていた。

近代になっても、橋のたもとの重要性は受け継がれ、大正前期、**日本橋**のたもとには、村井銀行、西川布団店をはじめ、西洋風の本格的な建築がずらっと並び、いかにも東京らしい水辺の広場を形づくった（図7）。震災復興事業においても、モダンな橋のたもとにとられる橋詰広場が重要なものとして位置づけられ、都心、下町に数多く実現し、緑の小広場のネットワークを生み出した。

近年、東京の水辺空間の復権とともに、隅田川やベイエリアに、水に開いた気分のいい広場がいくつか生まれているのが注目される。その一つ、江戸の橋のたもとの広場を現代に蘇らせたかのように、歩行者専用の「**桜橋**」のたもとに親水広場ができている（図8）。花見の時期に行われる早

Nihonbashi Nishikawa lined up at the foot of the Nihon Bridge, and the space there formed a very typical Tokyo waterside public space (Fig.7). Public spaces at a foot of modern bridges were valued as important in the earthquake reconstruction project. A lot of these spaces were established at city centers and downtown areas and produced networks of small green public spaces.

Restoration of waterside spaces in Japan as well as the establishment of several relaxing public spaces at along the Sumida River and bay area has attracted attention in recent years. One of these public spaces is a space close to water at the foot of the pedestrian "Sakura Bridge", which is a modern version of a public space at the foot of a bridge in the Edo period (Fig.8). On the day of the regatta Waseda vs Keio during the cherry blossom viewing season, cheering groups from both Waseda University and Keio University are stationed here and the celebratory atmosphere reaches a climax.

I would like to elaborate on waterside public spaces. Rivers in Japan are rapid streams and they are fated to bring flood damage. River beaches which are unstable grounds are one of the most distinctive spaces in Japan, and as I mentioned above, free urban activities tend to occur here. Theatrical performances were established here and restaurants with Kawadoko (riverbed seating spaces) were also established on long ago. We can say that the scenery of Kawadoko lining up along the Kamo River in Kyoto is a typical Japanese style public space (Fig. 9).

This is how the spaces along rivers were used practical-

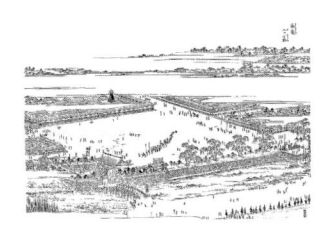

図6
筋違八ツ小路（江戸名所図会より）
Figure 6
"Sujikai Yatsukoji" from "Edo Meisho Zue (Guide to famous Edo sites)".

図7
大正前期頃の日本橋（古絵葉書より）
Figure 7
Nihon Bridge, In the early Taisho period (from old postcard)

図8
桜橋（隅田川）
Figure 8
Sakura Bridge (Sumidagawa riv.)

慶レガッタの日には、ここに両校の応援団が陣取り、祝祭気分は最高潮に達する。

　水辺の広場をもう少し考えてみたい。そもそも我が国の河川は急流で、水害をもたらす宿命にあった。不安定な土地である河原は、日本における一つの特徴的な空間で、すでに前述したように、そこに自由な都市的活動が芽生える傾向が見られた。芝居も河原で生まれたし、川床をもつ料理屋なども早い時代からそこに登場してきた。京都の**鴨川**に面して川床が並ぶ光景は、日本的な広場の典型といえよう（図9）。

　こうして川沿いの空間は、人々によって積極的に活用されてきた。しかし、近代の治水の思想と技術が広がるにつれ、河川改修事業で水際には高い堤防がつくられ、河川区域内からは、民間の営業活動は一切閉め出された。川沿いの空間に見られた日本的な広場性は失われた。

だが、河川法が改正され、治水に加え環境を考慮し、さらに河川の市民・住民による活用を求める方向が打ち出され、国交省も規制緩和の方向に大きく舵をとりはじめた。こうした動きを背景に、広島市で、我が国で初めての**川沿いのオープンカフェ**（図10）が2007年より本格的に定着・実現したのである。それを皮切りに、名古屋の**堀川**（図11）、さらには、2013年、東京台東区浅草の**隅田公園**に、川沿いのオープンカフェが2つ登場し（図12）、日本らしい水辺の広場が再び生まれている。

近代・現代の主役となった駅前の広場

　橋詰広場に話を戻そう。水の都市から陸の都市に転換した近代の東京では、橋のたもとの代わりに、駅前が都市における人の流れの結節点とな

ly and actively by Japanese people. However, as techniques of modern flood control are spread out, high embankments have been built close to the water in river improvement projects and private sales activities were banned at river areas; characteristics of Japanese style public spaces along rivers have thus disappeared.

　However, the River Act was revised with consideration taken not only of flood control but also of the surrounds. Residents have also requested the practical use of rivers; the Ministry of Land, Infrastructure, Transport and Tourism has therefore taken to deregulation. At this movement, the first proper open cafe along a river (Fig.10) was established in Hiroshima City in 2007. Since starting with this cafe, another open cafe along the Hori River (Fig.11) in Nagoya

and two more open cafes in the Sumida Park in Asakusa, Taito-ku, Tokyo in 2013 were established; Japanese style waterside public spaces are thus being reestablished (Fig.12).

Station Squares as Leading Roles in Modern Times

I would like to go back to the subject of public space at foot of bridge, "Replacing" the role of the foot of a bridge, spaces in front of train stations have become a node point of the flow of people in modern times in Tokyo. Coinciding with Tokyo's changing from a "water city" to a "land city" it is interesting to see that public spaces have followed this change in node points. The train station is the face of a town. It is also a mark to know where we are in metropolis

図9
鴨川　納涼床（京都市）
Figure 9
Kamogawa riv., riverbed seating spaces (Kyoto city)

図10
川沿いのオープンカフェ（広島市）
Figure 10
Open cafe along a river (Hiroshima city)

図11
堀川に架かる納屋橋（名古屋市）
Figure 11
Nayabashi bridge over Horikawa riv.

り、そこに広場が誕生したというのも、興味深い。駅は街の顔でもある。複雑な形態の巨大都市にあって、自分の位置を知る目印でもある。渋谷の**ハチ公の銅像のある広場**（図13）などは、近代日本の生んだ典型的な広場といえよう。

新橋駅烏森口の駅前広場（図14）は、東京の中でも、最もイタリアの広場に近い存在かもしれない。待ち合わせに使う人が多いが、何の目的もなくとも、いつまでも居られそうないかにも都市的な、自由度の高い広場なのだ。周囲のビルのスクリーンに映される映像を見ていても時間が過ぎる。女性の比率が少ない男ばかりの空間であるのも、南イタリアの広場とよく似ている。

駅前も車の量が少ないと、のどかにくつろげる雰囲気のある広場が可能になる。真ん中に噴水を置き、ベンチと植え込みで囲った**阿佐ヶ谷駅前広場**などは、人々に開放されうまく機能している駅前広場の一つだ（126頁）。

現在、経済事情の悪化のせいか、全国的に見て、既成市街地の再開発や再生が頭打ちなのに対し、JRの駅前広場の改造事業が各地で成果をあげている。中でも目を引くのは、建築家・西沢立衛の設計になる**熊本駅東口駅前広場**（56頁）で、"白い雲"をイメージしたという巨大な屋根ができて、公園のように人々が集い憩えるパブリックな空間が創造され、駅前広場の概念を大きく変えた。

姫路駅北駅前に実現したばかりの広場もなかなか良い（74頁）。姫路の駅前は、世界遺産・国宝姫路城を擁する城下町にふさわしく、「城を望み、時を感じ人が交流するおもてなし広場」というデザインコンセプトの下、国内最大級の駅前広場へと生まれ変わった。商業施設と組み合わせた新駅ビルの2階部分の中央コンコースの端のデッ

within a hugely complex system. We can say that a public space with the statue of Hachiko in Shibuya (Fig.13) is a typical public space in the modern times of Japan.

The Shimbashi Station Karasumori Exit station square (Fig.14) might be considered the most similar Japanese public space to those in Italy. Although a lot of people use the station square as a meeting spot, it is also a highly flexible urban style public space, where we feel that we can stay as long as we want to without any particular purpose. Time flies even when we are just watching images on the screens of surrounding buildings. The ratio of women is less than the ratio of men in the station square and it is very similar to piazzas in Southern Italy.

Station squares can have a relaxing atmosphere if the traffic is not bad. For example, the Asagaya Station square, which has a water fountain in the middle and is surrounded by benches and plantations, is a station square that is functioning well by being opened up for people (p. 126).

Although redevelopment or regeneration of built-up areas has past its peak due to worsening economic conditions, reconstruction projects of Japan Railways (JR) station squares have given positive results in various districts. One eye-catching station square is the Kumamoto Station east exit station square (p. 56) designed by architect, Ryue Nishizawa. It has a big roof representing "white clouds" and forms a public space where people can gather and relax. It has dramatically changed the concept of the station square.

図12
隅田公園オープンカフェ（東京都台東区）
Figure 12
Open cafes in the Sumida Park (Taito-ku, Tokyo)

図13
ハチ公前広場（東京都渋谷区）
Figure 13
Hachiko in Shibuya (Shibuya-ku, Tokyo)

図14
新橋駅前広場（東京都港区）
Figure 14
The Shimbashi Station square (Minato-ku, Tokyo)

キに立つと、北に延びる城下町の空間軸（大手前通り）の先に、姫路城が望める。その駅前に、格好いい広場が実現している。手前は大きく掘り込んだサンクンガーデンで、池と噴水のある人工的なオアシスのイメージ。その北側の地上には、気持ち良い芝生広場が大きくとられ、昼も晩も多くの市民を集めている。都市計画の醍醐味を感じさせる素晴らしい広場だ。

現代が生んだ新たなカテゴリーの広場

これまで、伝統の系譜にのる日本らしい広場といえる場所を中心に見てきたが、ここからは、過去には存在しなかった現代日本ならではの新しいタイプの広場にさらに目を向けていこう。

1980年代後半以後、大きな再開発とともに登場した広場をまず見よう。東京では、好景気に沸いたバブル経済の時代、幾つもの再開発計画が生まれ、90年代に入ってそれらが実現した。そもそも日本は、ヨーロッパの都市に比べ、歩行者空間化では大きく遅れをとってきた。**銀座の中央通り**（130頁）、新宿通りなど、歩行者天国で成功し、人気を集める場所も幾つかあるが、むしろ比較的最近、歩行者空間が廃止されたものも多い。

まるでそれを補うように、再開発で実現したプライベートな敷地の空間を活用して、広大なその土地に車の入らない人間に開放された広場や広々とした街路をつくり、カフェテラスを設け、賑わいのある魅力的な空間を実現した例が幾つも登場している。

新宿駅南口から代々木方面に広がる新宿サザンテラス（1998年4月にオープン）は、小田急線の路線上に人工地盤を築き、遊歩道をイメージして店舗を配置した。夜は下を通るJRの電車が、

The newly established public space in front of the north side of Himeji Station is notable too (p.74). The space in front of Himeji Station is appropriate for the World Heritage and National Treasure, Himeji Castle. It was reestablished as one of the largest station squares under a design concept of "a hospitable public space where we can overlook the castle, appreciate the passing of time and interact with others". We can overlook Himeji Castle at the north end of a space axis (Otemae-dori Street) of the castle town from a deck located at an edge of a central concourse on the 2nd floor of the new station building. This is combined with commercial facilities. A picturesque public space emerges in front of the station. A sunken garden, which was dug in widely and represents an artificial oasis with ponds and water fountains, is at the near side. A nice feeling of big public space with grass is located on the north side of the sunken garden on the ground and attracts a lot of people both during the day and at night time. This public space can shows us the real pleasures gained from good urban planning.

Public spaces in a New Category in Modern Times

I have talked mainly about Japanese style public spaces in the context of the traditional linage of public spaces so far. Let's turn our attention to a new type of public space which has only existed in modern times in Japan.

Firstly, there are those public spaces that arose from big redevelopments after the late 1980's. In Tokyo, several redevelopment projects were planned during the booming "bubble economy" period but were not completed until the 1990's. Compared to Europe, Japan has significantly fallen behind in creating pedestrian spaces. Although there are some popular places with a pedestrian zone, such as Chuo-dori Street in Ginza (p. 130) and Shinjuku-dori Street, rather, the development of pedestrian spaces has, relatively speaking, declined in recent years.

Several attractive spaces with lively atmospheres were created with large public spaces specifically for pedestrians, with wide streets and cafe terraces by using spaces belonging to private properties. This is a practical approach and has helped boost public spaces for pedestrians.

At Shinjuku Southern Terrace (opened in April, 1998) starting from Shinjuku Station south exit towards Yoyogi area, shops were arranged to look like a promenade walk

光の川の流れのように見えて、ロマンチックな雰囲気を楽しむカップルも多い。一方、品川駅東口にあった旧国鉄の操車場跡地にできた**品川インターシティ**の広場（図15）は、ランドスケープの洒落たデザインがなされ、オフィスで働く人々の憩いの場になっている。ただ、最も有効に使っているのは、もっぱら喫煙者のようにも見える。せっかくの広場が、喫煙者のたまり場と化している所も少なくない。

ビール工場の跡地に生まれた**恵比寿ガーデンプレイス**（図16）には、広場が意欲的に創られた。商業施設が充実し、都心に住む人々が増えたこともあって、ベビーカーを押す母親、高齢者も含む周辺住民もベンチに座り、生活感が生まれてきている。スクリーンを設け、広場でのイベントも行われる。2003年に開業した六本木ヒルズは、超高層の森タワーの足下に、一部に屋根を持つ屋外型イベントスペースとしての広場、**六本木ヒルズアリーナ**（図17）を設け、円形ステージでのライブイベントから広場全体を使ったパフォーマンスまで、多彩な催し物を行い、都心での新たな広場のあり方を示している。

近年の広場で最も成功していると思うのが、丸の内に復元された三菱一号館と超高層の丸の内パークビルの間に実現された**三菱一号館広場**だ（2009年9月オープン、112頁）。オフィス機能しかなかった丸の内に、三菱地所の新戦略で商業・文化の魅力的な機能が導入されてきたが、その成果がここに象徴的に表れている。煉瓦の復元された歴史的建造物の背後にある緑溢れ、ゆったりと時間が流れる公園のようなこの広場には、カフェなどの商業機能が入って、ほどよい賑わいもあり、憩いの空間として、とりわけ女性たちの人気を集めている。コーナーからさり気なく入る

along an artificial ground on the Odakyu train line. At night time, JR trains going under the terrace look like a light river flow and a lot of couples enjoy this romantic view. In another example, a public space at Shinagawa Intercity (Fig.15), which was created on a site of demolished marshaling yard of old Japanese National Railways at Shinagawa Station east exit, has a stylish landscape design and has become a relaxing place for office workers. However, it does seem to attract more smokers than anyone else. Public spaces are precious, but unfortunately, many of them have become a favored gathering spot for smokers.

In Yebisu Garden Place (Fig.16), constructed on the site of a demolished brewery, a public space was created intentionally. In the Yebisu area, more commercial facilities have been constructed and the population of the city center has increased. Therefore, we start to get the feeling that people are actually living in that area by watching mothers pushing prams and residents, including old people, sitting on benches. Some events using installed screens are also held at the public space. Roppongi Hills, opened in 2003, has a public space called Roppongi Hills Arena with some covered areas as an outdoor event space at the foot of the high-rise Mori Tower. Various events from live concerts events in the circular stage to performances using the whole area of the public space are held and it shows the forming of a new state of public space in city centers (Fig.17).

I believe that the most successful recent public space

図15
品川インターシティ（東京都港区・品川区）
Figure 15
SHINAGAWA INTERCITY
(Minato-ku, Shinagawa-ku, Tokyo)

図16
恵比寿ガーデンプレイス（東京都渋谷区）
Figure 16
Yebisu Garden Place (Shibuya-ku, Tokyo)

図17
六本木ヒルズアリーナ（東京都港区）
Figure 17
Roppongi Hills Arena (Minato-ku, Tokyo)

と、目の前にぱっと緑の広場が登場するという演出が、実に生きている。現代日本ならではの都心型広場となっている。

このように1990年代以後、広場を積極的に取り込んだ多くの再開発が実現した。その傾向を見ると、住宅地が近いと「生活感のある広場」が、繁華街を背景にすると「エンターテインメント性のある広場」が、駅近くのビジネス空間には「オフィス街の広場」が、かつてなかった方法で意欲的につくられている。

目立つ事例としてはこうした大都市の民間の商業施設、オフィスの複合形が多いなか、地方都市において、公共施設が意欲的に広場空間を創出した異色の建築作品が、隈研吾の設計で2012年に完成した**アオーレ長岡**（18頁）だ。中心市街地の衰退を食い止めるべく、市役所を駅近くの都心に戻し、その内部に、あらゆる世代の多様で自発的な活動を実現する場として、また、誰もが憩い集う「市民交流の拠点」として、大屋根の下の広場「ナカドマ」がつくられたのだ。周辺市街地との共存を目指した新しいコンセプトの市庁舎として注目されている。

2004年開館と少しだけ時期は遡るが、**金沢21世紀美術館**（54頁）（設計者：SANAA）も、公共側が生み出した新たな広場としての性格をもつ施設といえよう。金沢城公園のすぐ近くの文化ゾーンに登場したガラス張りの円形の建物で、どこからでもアクセスでき、誰もがいつでも立ち寄ることができ、さまざまな出会いの「場」となるような美術館を目指している。

水辺に賑わいを生む広場

日本にも、こうして本格的な広場が登場してき

is Ichigokan Hiroba (opened in September, 2009, p. 112) between Mitsubishi Ichigokan and the high-rise Marunouchi Park Building. Commercial and cultural attractive functions have been introduced in the Marunouchi area, which used to have only office functions. This is the result of a new strategy of Mitsubishi Estate and its result is symbolically represented here. This public space located at the back of the renovated historical brick building is sylvan and feels like a park where you appreciate time passing slowly. It has commercial functions including cafes and a good lively atmosphere. It is popular especially with women as a relaxing place. The feeling that a green public space suddenly opens up to us as we come in from one corner is especially effective. This urban style public space can only be possible in today's Japan.

As I mentioned just now, a lot of redevelopments actively taking public spaces in were carried out after the 1990's. There is a tendency to provide "public spaces where we can feel that people are actually living in the area" near residential areas, "public spaces with entertainment characteristics" near busy downtown districts and "office town public spaces" in commercial and business districts near stations. These public spaces are inspired by methods which were not available to us previously.

The most of distinguishable examples are the private commercial facilities found at office complexes. However, a notable architectural work that was create by the public sector as a public space within a provisional city is the Nagaoka City Hall Aore (p. 18) which was completed in 2012 and designed by Kengo Kuma. The city hall was moved back to the city center near a station to stop the decay of the city center. In the city hall, a public space under a big roof called "Nakadoma" was created as a place to support the various self-initiated activities of all generations and as a "base of residents' interaction" in which anyone can gather. It is getting attention as a city hall with the novel concept of co-existence of urban and suburban areas.

21st Century Museum of Contemporary Art, Kanazawa (p.54) (designer: SANAA), which was opened in 2004(it is a little bit older than other facilities), is a facility with characteristics of new public space created by the public sector. It is a circular glazed building in a cultural zone near the Kanazawa Castle Park. It aims to be a "place" for various encounters, which anyone can access and stop by from anywhere, anytime.

たと感じられるようになったのは、1990年代以後である。晴海トリトンスクエア（2001年竣工、図18）も、そう思わせる好例だ。朝潮運河に面する水辺にあり、海に囲まれた街である晴海をギリシア神話の海の神であるトリトンと掛け合わせている。3棟のタワー・オフィスビル、複合商業施設に加え、高層の住居棟を背後にもって、運河の水に開く半円形の広場がつくられた。オープン当時は有名ブランドの店も入り話題を呼んだが、時間とともにむしろ地元の人々を対象とする商業空間になり、広場も周辺住民が集まる生活感のある空間に成熟しつつある。

月島と向かい合う位置に、これより少し前に登場した聖路加ガーデン（竣工1994年）も水辺の広場の好例で、2棟の超高層ビルの足下に、スーパー堤防の方式で緩勾配の階段で水際まで降りられる親水テラスを設けている（137頁）。水辺で昼食を楽しむ勤め人、犬の散歩で水辺のプロムナードを歩く人々の姿が多く、花見の際も大いに賑わう広場となっている。

水辺に実現した広場は他にもいくつかある。品川の天王洲の京浜運河に面した一角にある、かつてのお台場の上に登場したシーフォート・スクエア（1992年竣工、図19）もその例の一つだ。ホテル・住宅・商店・オフィスが入ったビル群のコンプレックスの足下に、水に開いた気持ちのよい広場がつくられている。

豊洲に2006年に登場したららぽーと豊洲（図20）は、旧石川島播磨重工業（現・IHI）の造船所のドックを保存活用して建設され、超高層マンション群と隣接している。定期便の水上バスが運行されていて、水の東京をアピールする広場となっている。

これらの再開発に伴って登場した広場だけでな

Public Spaces with a Lively Atmosphere along the Waterside

After the 1990's, we started to feel that fully-fledged public spaces had been created. The Harumi Triton Square (completed in 2001, Fig.18) is a good example of these public spaces. It is at the waterside of the Asashio canal, and Harumi, which is a town surrounded by the ocean and is compared to Triton, mythological Greek god of the sea. A semicircular public space, open for the water of the canal, which has 3 tower office buildings and a shopping mall as well as a high-rise residential building at the back, was created. When the Harumi Triton Square was opened, it enjoyed popularity because of famous brand shops. However, over time it changed into a commercial space for local people, and the public space is getting more mature as a space with a "lived-in" feel as it has more local residents now.

The St Luke's Garden (completed in 1994, p. 137), which was constructed a little bit before the Harumi Triton Square, located opposite the Tsukishima train station is also a good example of public space by the riverside. It has a terrace close to the water at the foot of 2 very high buildings on a large river embankment and we can walk down on gently sloping stairs from the buildings to waterside. We can see office workers enjoying their lunch at waterside and people walking their dog on waterside promenade. It also has a big lively atmosphere during the cherry blossom viewing season.

図18
晴海トリトンスクエア（東京都中央区）
Figure 18
Harumi Triton Square (Chuo-ku, Tokyo)

図19
シーフォート・スクエア（東京都品川区）
Figure 19
Seafort Square (Shinagawa-ku, Tokyo)

図20
ららぽーと豊洲（東京都江東区）
Figure 20
LaLaport TOYOSU (Koto-ku, Tokyo)

く、既存の水の空間を活かして広場に転換させる試みも見られる。法政大学の私の研究室では、外濠の水上にある**カナルカフェ**(134頁)と組んで、2007年頃から、夏の夜に、水上コンサートを実施して、好評を博してきた。水辺のデッキの先端にステージを設け、その手前と水上のボートから音楽を楽しむ趣向であり、東京の都心にいることを忘れ、豊かな気分を満喫できる。まさに水上に出現する広場といえる。

　ところが、困ったことが一つある。音量を上げ過ぎると、周辺のマンション住民からすぐにクレイムがつくのだ。最近、日本の都心に住む人たちは、下町のかつての粋な文化を忘れて個人主義を強め、運動会の子供の歓声、船の汽笛、花火、野外コンサートの音などに過度に敏感になり、クレイムを付けたがるのである。個人主義が発達しているといわれるヨーロッパでも考えられない現象

が生まれている、といわざるを得ない。都心に住む楽しさを皆で共有しようというおおらかさがないと、本当の広場は成立しない。

屋上広場の出現

　最近の広場を巡る新しい動きとして、既存のビルの屋上に緑に溢れる広場が続々と登場しているのも注目される。

　デパートの屋上を有効に使う工夫は、昭和初期の震災復興時から行われてきた。隅田川に面して颯爽と登場したモダン・デザインの松屋浅草店の屋上には、大きな遊園地ができ、空を行く「航空艇」と呼ばれるロープウェイが設けられた。戦後も各デパートの屋上には、ちょっとした遊園地、ステージのあるイベント空間などがあった。屋上ビアガーデンも長い間、人気者だった。だが、都

There are several other public spaces along the waterside worth noting. One example is the Seafort Square (completed in 1992, Fig.19) located at one corner facing the Keihin canal in Tennozu, Shinagawa, which was established at the site of former Odaiba. A nice feeling public space by the water was created at the foot of a complex of buildings with hotels, residential areas, businesses and offices.

LaLaport TOYOSU (Fig.20), which appeared in Toyosu in 2006, was constructed by preserving and utilizing the shipyard and dock belonging to Ishikawajima-Harima Heavy Industries (current IHI Corporation) and it is next to a group of very high apartments. A regular water-bus is running and a public space was created to give the sense of Tokyo as a water city.

Public spaces by the water in Japan are not only the result of redeveloping land areas, but in fact, have also been created using existing water space. My Laboratory at Hosei University and CANAL CAFE on water next to an outer moat have been working together to hold water concerts at night in summer since 2007 and it has been gaining popularity (p. 134). A stage was created at the point of a waterside deck and we can enjoy music in front of the stage or on a boat. We lose the sense that we are in the middle of Tokyo and we are able to fully appreciate the moment. It is a public space arising from the water.

However, Japanese cities are suffering from a peculiar backlash from events, such as this water concert, held in public spaces. If we turn volume up too much, residents of surrounding apartments complain. Nowadays, residents in the city centers of Japan forget the old downtown culture and have stronger individualism; they tend to be too sensitive and complain about noises, such as the cheers of children on sports days, whistles of ships, fireworks and noises coming from live outdoor shows. While Japanese people feel that individualism is a European trait, there is in fact a contradiction here because, Europeans themselves are able to cope with the noise arising from public spaces. Real public spaces will succeed in Japan unless we have generosity to share the joys of living in city centers with everyone.

Emergence of Rooftop Public Spaces

As a new movement of public spaces, the emergence of green rooftop opens spaces gaining attention.

The idea of making use of the rooftop spaces of department stores was formed in Japan during reconstruction

市に人を集める場所がどんどん生まれると、古くさい屋上の空間はまったく忘れられた存在となっていった。

しかし、時代が巡り、既存のデパート、商業建築のリノベーションにともない、再び屋上空間に光が当たってきた。環境、エコロジーへの関心が高まり、都市緑化などの社会的な貢献も考慮し、屋上に従来とはまったく違う緑溢れる憩いの広場が生まれているのだ。

その走りともいえるのが、**伊勢丹新宿店本館**の緑に包まれたアイ・ガーデンと呼ばれる屋上広場だ（図21）。1933年竣工のアールデコの建築を大切にし、幾度もの改修工事を行ってきたもので、屋上緑化は2006年に実施された。単なるデパートの屋上集客空間とすることなく、今日的なテーマ性をもって、都市の公共性に鑑み、芝生広場や四季をテーマとした回遊式庭園など、積極的な都市の緑化と公共の開かれた空間となっている。「都会の喧噪のなかでのアーバンオアシス」として、気持ちのクールダウンできる空間を目指しているという。ベビーカーを押す子供連れの若い母親が多いのが特徴で、芝生で走り回る大勢の子供の姿が印象的だ。親たちもベビーカーを近くに置き、子供たちと楽しんで遊んでいる。一つある小さな滑り台も人気で、庭園を抜ける小径にも走っている子供の姿がある。都心のマンションに住む人口が増えてきたことが、このような現象を生むようになった最大の理由なのであろう。庭のない都心の集合住宅に住めば、広場のような公共的な空間が必要になるのは、ヨーロッパが長らく経験してきたことで、近年の東京にも、同じ土壌が成立してきたと考えられる。

道をはさんですぐ近くにある**新宿マルイ本館**は、館全体を、職場・家庭に次ぐ第3の場所（サ

that was carried in the wake of an earthquake disaster in the early Showa period. A large amusement park was created on a roof of the modern Matsuya Asakusa department store, which arose dashingly along the Sumida River. A ropeway called the "flying boat" in the sky was installed. Small amusement parks, event spaces with stages and the like were created on the roof of each department store in the postwar era. Rooftop beer gardens were also popular for a long time but then, as gathering places for people started to increase, people started to forget about the existence of old fashioned rooftop spaces.

However it appears that the light has started to shine on rooftop spaces once again with renovation of existing department stores and new commercial architectures.

図21
伊勢丹新宿店本館　屋上庭園「アイ・ガーデン」（東京都新宿区）
Figure 21
Isetan Shinjuku Main Building , "I Garden" (Shinjuku-ku, Tokyo)

Relaxing public spaces full of green, which are totally different from the existing ones, have emerged as a consequence of increasing of concerns about the environment and ecology and with social contributions including urban tree-plantation.

The first public space in this new movement is the "I Garden" which is a rooftop public space with full of green on the Isetan Shinjuku Main Building (Fig.21). Shinjuku Isetan had several renovations but still retains its art deco style of architecture from its 1933 construction. Rooftop greening was carried out in 2006. A modern theme of "Urban Public Nature" with active urban greening was adopted creating a public space where the public can utilize grass lawn, a circuit style garden changing with the seasons and the like. The aim of this space is to become an "urban oasis within a bustling big city", where people can feel calmer. The space is characterised by the presence of young mothers pushing prams and it is impressive that a lot of children are running around on its lawn. Parents put their prams aside and have fun with their children. One slide within the space is popular and children are also seen running along the path to the garden. The biggest reason of this phenomenon is most likely the increase in the population living in apartments in the city center. In recent times, Tokyo is starting to resemble

ードプレイス）として、客に憩いの場を提供することをコンセプトに、緑の空間を多く確保し、屋上には、イングリッシュガーデンを実現した（2009年開設、106頁）。「木漏れ日の庭」「早春の庭」「門出の庭」「実りの庭」「薔薇の庭」の5つの個性あるスペースが配置され、歩く人々は季節の花の香りを楽しみ、水の音で癒される。新たな感性、ライフスタイルを開拓する女性がターゲットで、ショッピングの合間に屋上に出ると、緑豊かな庭園と青い空が広がり、都会とは思えぬ安らぎを覚える。

　緑つながりの広場ということで、最後に少し、東京の郊外にも目を向けておこう。「水の郷」日野にある「せせらぎ農園」は、素掘りの用水路が流れる農地に、地域の約200世帯から出る生ごみを肥料として有効活用し、無農薬・無化学肥料で野菜や花を育てているコミュニティガーデンである。約650坪の畑には、素掘りの用水路が流れ、かつて水田だったころの面影を残し、現在は子供たちが水遊びをする場所にもなっている。地域の交流を支えるコミュニティガーデンで、地主と一緒に市民が運営している。地域の小学校などの環境学習の場としても利用されている。農園の活動に参加する人々は、自分たちで育てた収穫物を皆で料理して、しばしば楽しんでいる。まさに、農業ゾーンに生まれた地域の人たちの交流の広場といえる。

結―「しなやかな広場」の時代へ―

　以上、現在の日本の都市にある広場を伝統的な系譜から読み、また新たに登場している今の日本らしい広場のあり方を紹介し、その成立背景を考えてきた。

Europe and we will need public spaces as public spaces if people live in apartment without gardens.

In Shinjuku Marui Main Building located at the opposite side of the street of the Shinjuku Isetan, a lot of green spaces and a rooftop English garden were created under the concept of providing a relaxing place for customers as a third place following work and family (opened in 2009, p. 106). It has 5 unique spaces: "a garden with sunlight filtering through the leaves of trees", "a garden in early spring", "a garden for departure", "a harvest garden" and "a rose garden". People can enjoy the smell of seasonal flowers and relax with sound of water while walking. The target of this space is women exploring their own lifestyles and when we go up to the rooftop whilst shopping, we can enjoy a change of scenery with a green garden blue sky above; we can feel peacefulness even in the middle of a big city.

Let's finally turn our attention to a particular suburb of Tokyo which warrants attention because it has become a space that is full of green. The "Seseragi Farm" is in Hino, or "Mizunosato" which has historically had a good record of water conservation and good town planning in relation to water use. The "Seseragi Farm" is a community garden which utilizes kitchen garbage from approximately 200 local families on the farmland with unsupported excavated canals and grows organic vegetables and flowers. The farm, with an approximate area of 2,148m^2, has unsupported excavated canals but still resembles a rice paddy. It is also a place for children now play. The landlord and residents run the community garden together, supporting the interactions of the local area. It is used also for environmental studies of the local elementary schools. People joining the farm activities sometimes enjoy cooking together with their harvests. Indeed, this is an urban space that has arisen from an agricultural zone for the interaction of local residents.

Conclusion
—To The Time of "Supple Public Spaces"—

I have discussed current public spaces in Japanese cities with reference to the traditional sense public spaces. I have also introduced the state of today's Japanese style public spaces and talked about the background of its formation. Through the discussion in this article, we can confirm that Japanese style public spaces have indeed been established and used in a variety of ways, refuting the suggestion that "public spaces did not develop in Japan". We also notice that the younger generation and women benefit from public spaces in a variety of ways. When establishing public

こう見てくると、かつて「日本には広場が発達しなかった」と言われたことが嘘のように、日本的な広場が多彩に生まれ、使われているのを確認できる。その広場の恩恵にあずかっているのが、女性や若い世代の人たちであることにも気づかされる。ライフスタイル、美意識、価値観が大きく変化してきたのは事実で、そのことが、広場を成立させる要因となっているに違いない。そして、近年の新傾向として、都心回帰の現象と呼応する新しいタイプの広場が生まれていることが、注目される。**六本木ヒルズ 66 プラザ**（84頁）にも、誕生したばかりの**虎ノ門ヒルズ オーバル広場**（114頁）にも、緑のなかを子供が大勢走り回る光景を見ることができるのだ。

イタリアのかつての広場のイメージは、建物で囲われた都心の堂々たる公共空間に、リタイアした男どもが昼間から集まり、人の輪を幾つもつくって立ち話をしている、といった光景だった。そうした重厚な広場の古典的なイメージとはずいぶん違う、女性のしなやかな感性で印象づけられる新たな広場が、今の日本にたくさんつくられてきているのに驚かされる。それは同時に、子供ばかりか高齢者にとっても居心地のいい、安らぎのある憩いの空間となりえる。今後、都心の集合住宅に住む人口が増えるほど、このような緑や水を取り込み、ランドスケープの魅力を加えた日本型の広場の需要が高まるに違いない。

日本の広場は、このようにいつの時代にも、人々の身体や心理にフィットした居心地のよい、そしてまたほどよい刺激のある場所に生まれ、使いこなされてきたことがよくわかる。広場の問題を考えるには、＜空間＞と＜人間＞の対話のあり方を深く探る必要があるのだ。

spaces one must factor in ever-changing lifestyles, sense of beauty and sense of values. Moreover, following recent trends, we must pay attention to a new type of public space resulting from the phenomenon of increasing populations in city centers. We can witness a lot of children running around the green spaces in the Roku-Roku Plaza of Roppongi Hills (p. 84) and in the brand new Oval Plaza of Toranomon Hills (p. 114).

The existing image of piazzas is the one that I described in Italy where retired men come to congregate in a majestic public space, surrounded by urban buildings and others come during the day to gather, standing and talking in circles. It is a pleasant surprise that a lot of new public spaces in Japan reflect the supple sensibility of women, which is totally different from profound classical public spaces. At the same time, these new spaces provide peacefulness not only for children but also for old people. In the future, the demand of Japanese style public spaces with landscape attractions, which incorporate greenery and water features, will increase as more people live in apartment houses in city centers.

Now, we can tell that Japanese public spaces have arisen and used in a place with appropriate stimulus anytime, which is suitable for people's body or state of mind and comfortable. We need to explore how the conversation between <space> and <human> works deeply in order to discuss about issues of public spaces.

日本の広場　事例集

Public Spaces in Japan

金沢21世紀美術館
21st Century Museum of Contemporary Art, Kanazawa

たまり場であり、パッセージでもある。
まちに開かれた公園のような美術館

Gathering spot as well as passage
Museum open to the city like a park

Design
妹島和世＋西沢立衛／SANAA
Kazuyo Sejima + Ryue Nishizawa / SANAA

DATA
Site: Kanazawa city, Ishikawa, Japan
Principal use: Museum
Total floor area: 17069㎡ (museum)
Completion: 9/2004
Structure: RC + S structure

多くを語るまでもなく、ここが現代建築の傑作の一つとして、国内のみならず世界にその名をとどろかす美術館であることは疑いない。「まちに開かれた公園のような美術館」という建築コンセプトのとおり、誰もがいつでも立ち寄ることができ、様々な出会いと交流が実現されている。

　館内は有料の展覧会ゾーンと無料の交流ゾーンに分かれ、館内外の至る所で、五感で楽しむことのできる、建物と一体化した作品が鑑賞できる。

　地域住民にとって、この場はある一面では応接間のようだという。遠方からの来訪者をもてなす場として、とりあえず連れて来る。せっかくだから見ていってよと、少し誇らしげな気分で、まるで自慢の自宅のように案内する。そう思わせられる場なのである。そしてある一面では、居間のようでもある。芝生でくつろぐ者、用もなくただベンチで寝そべる者、雨天時に児童館のように子どもを遊ばせる母親。「帰宅時間、ちょうど明るいから通りたくなる」と話す勤め人もいた（交流ゾーンは22時まで開館）。ここはもはや金沢の街にとって欠かすことのできない一片(ピース)なのである。

As one of modern architectural masterpieces, there is no doubt that the name of the 21st Century Museum of Contemporary Art, Kanazawa is spread throughout the world as well as in Japan. Its architectural concept was a "Museum open to the city like a park". Per this concept, anyone can drop in to the museum anytime and various encounters and interactions are achieved there.

　The inside of the building is divided into a paid admission exhibition zone and a free admission interaction zone. You can enjoy art works integrated with the building using the five senses everywhere in the building.

　One aspect of this place is that it serves as a drawing room for its local residents. You can bring visitors from a distant place to this museum to entertain them. You say "Since you are here, why don't you have a look?", and proudly show around the museum as might your own home. This is how we feel when we are there.

　Another aspect of this place is that it serves as a living room. A person relaxes on the lawn. A person sprawls on a bench without anything to do and a mother lets her child play as though it is a children's house when it rains. Also, one office worker said "I would feel like going through the museum on the way home because it is brightly lit exactly when I go home." (the interaction zone is open until 22 o'clock). The town of Kanazawa would not be complete without this museum now.

日差しを和らげる雲形の「日傘」
A cloud-shaped "parasol" to ease the sunlight

Design
西沢立衛建築設計事務所
Office of Ryue Nishizawa

DATA
Site: Kumamoto city, Japan
Principal use: Station square
Total floor area: 907.31㎡
Completion: 10/2010
Structure: S + RC structure
Contractor: TEKKEN CORPORATION

熊本駅東口駅前広場（暫定形）
Kumamoto Station east exit station square

　熊本駅東口駅前広場暫定計画である。
　この広場の計画は２段階式で、第１段階としては平成23年の新幹線開業時に合わせた今回の計画部分（暫定形）がある。その後、平成30年竣工予定の新駅舎建替事業に伴って駅前広場が拡大される予定で、その際に再整備される広場全体を「完成形」と呼ぼう。設計者はその暫定形から完成形に至るまでの、都市空間としての時間的な連続性、調和性が重要と考えた。
　具体的な提案は「薄く軽やかな雲形の屋根を複数浮かべる」というものである。暫定形でまず一つ屋根が作られ、完成形においてさらにいくつかの屋根が作られる。屋根は、カーブしながら車や人、路面電車などの諸動線の流れから導き出されたカーブ曲率をおのおの持ち、柔らかくカーブし

This is a temporary plan of Kumamoto Station east exit station square.

The plan for the station square has two stages and the first stage is this temporary plan fitted at the time of a new bullet train operation commencing in 2011 (temporary version). After that, there is a place to expand the station square in a reconstruction project for a new station building scheduled to be completed in 2018. The station square will again be redeveloped and this is referred to as the "completed version". The designer thought that it is important to achieve temporal continuity and harmony in the urban spaces throughout the planning of the temporary version and the completed version.

The detailed proposal was "to create a number of floating thin and light roof with a cloud shape. First

ながら雲の群れのように空中に浮かび、熊本の強い日差しから人々を守る大きな日傘となる。構造としては、キャンチレバーのスチール柱群がRCのフラットスラブを支えるという、シンプルな計画が採用された。

　そこには、壁というものがない。誰でもどちらからでも気軽に訪れ、通り抜けることができるような、高い開放性と透明感を併せ持つ。

　駅前に集中する諸動線の処理だけに終始する交通空間ではなく、緑と太陽、日向と日陰から成る、人々と街に開かれた公共空間が作り出されている。

roof was created for the temporary version and some more roofs will be created for the completed version. Each roof has its own curvature derived from several flow lines of cars, people, trams and the like, and the roofs float in air like a group of clouds while curving gently and form a large parasol to protect people from Kumamoto's strong sunlight. For the structure, a simple plan comprising a group of cantilevered steel pillars, supported by a flat slab made from Reinforced Concrete (RC) was adopted.

　No wall can be found there. The station square has both high openness and a sense of translucence by giving you a feeling that anyone can visit and go through freely from all directions.

　The station square does not produce a transportation space only for dealing with the several flow lines converging to the station front area. Instead, it produces a public space open for people and the town, which was made up from green and sun, and sunshine and shade.

東急プラザ 表参道原宿「おもはらの森」
TOKYU PLAZA OMOTESANDO HARAJUKU, "OMOHARA-NO-MORI"

街並みに浮かぶ
建築と樹木の融合体

A fusion of architecture and trees floating in a townscape

Design
中村拓志& NAP建築設計事務所
Hiroshi Nakamura & NAP Co., Ltd.

DATA
Site: Shibuya-ku, Tokyo, Japan
Principal use: Commercial architecture
Total floor area: 11,852㎡
Completion: 3/2012
Structure: SRC + RC + S structure
Structure design, Contractor: Takenaka Corporation

最新鋭の建築が建ち並び、ケヤキ並木の枝葉が木漏れ日を作りなす。設計者はこの建築空間と樹下空間の一体化した環境こそが「東京・表参道」特有の空間であると考えて、その特徴を最大限取り込んだ環境形式が提案された。表参道の街並み上空に建物と樹木が一体となったボリュームを浮かせ、木漏れ日の空間を段階的に建物上部へ連続させたのである。これは鳥や蝶などが屋上の森に昇ってきやすいだけでなく、人々にとっても、屋上庭園がこの建物を訪れる一つのきっかけになり、商業的には建物最上階から下階へと客の流れをつくるシャワー効果を生み出すのである。

　問題となったのは、3階以上への客の誘引であるという。昇降時間の長さを少しでも短く感じられるよう、エスカレーターは「万華鏡の中のような」体験型空間とされた（61頁写真）。着飾った人々が万華鏡の中のカラフルなパーツのように反射を繰り返すことで、最先端の街を訪れているという人々の高揚感が更に増幅される。ミラーによって間口が実際より広く見える上に、反射によって人も倍増して見えるため、行列効果によって道行く人々を誘う。エスカレーターを昇ると大きなアトリウムが現れ、屋上につながるトップライトから木漏れ日が落ちてきて、表参道らしい買い物体験が始まる、というストーリーが描かれている。

　建物のなかで一番日当たりが良く、かつ交差点のどこからでも見える外周部と屋上には、樹木とテラスがレイアウトされた。その結果、室内は木々の枝葉によってえぐられ、屋上広場は根鉢の高さのぶん隆起している。しかし逆にこの不自由さが、人を樹木と密接な関係にさせているといえる。

　屋上は、根鉢の段差を多角形のステップで均すことで、すり鉢状の広場となった。これは階段や椅子、テーブルとしても使える多義的な形態で、使う人が身体的に建築と対話するきっかけになる。訪れた人は自然とすり鉢空間の底部に体を向け、中心を見つめるように座っていく。木漏れ日の下で小さな無数の入隅（いりすみ）に人々がはまって、視線の先を共有するのである。隣の人々と対面する緊張感がほどけて、ふと気づけば皆が一緒の輪になっている。こうした、人々の「ゆるやかな一体感」を生み出す空間演出力こそ、現代の商業施設に求められることなのかもしれない。

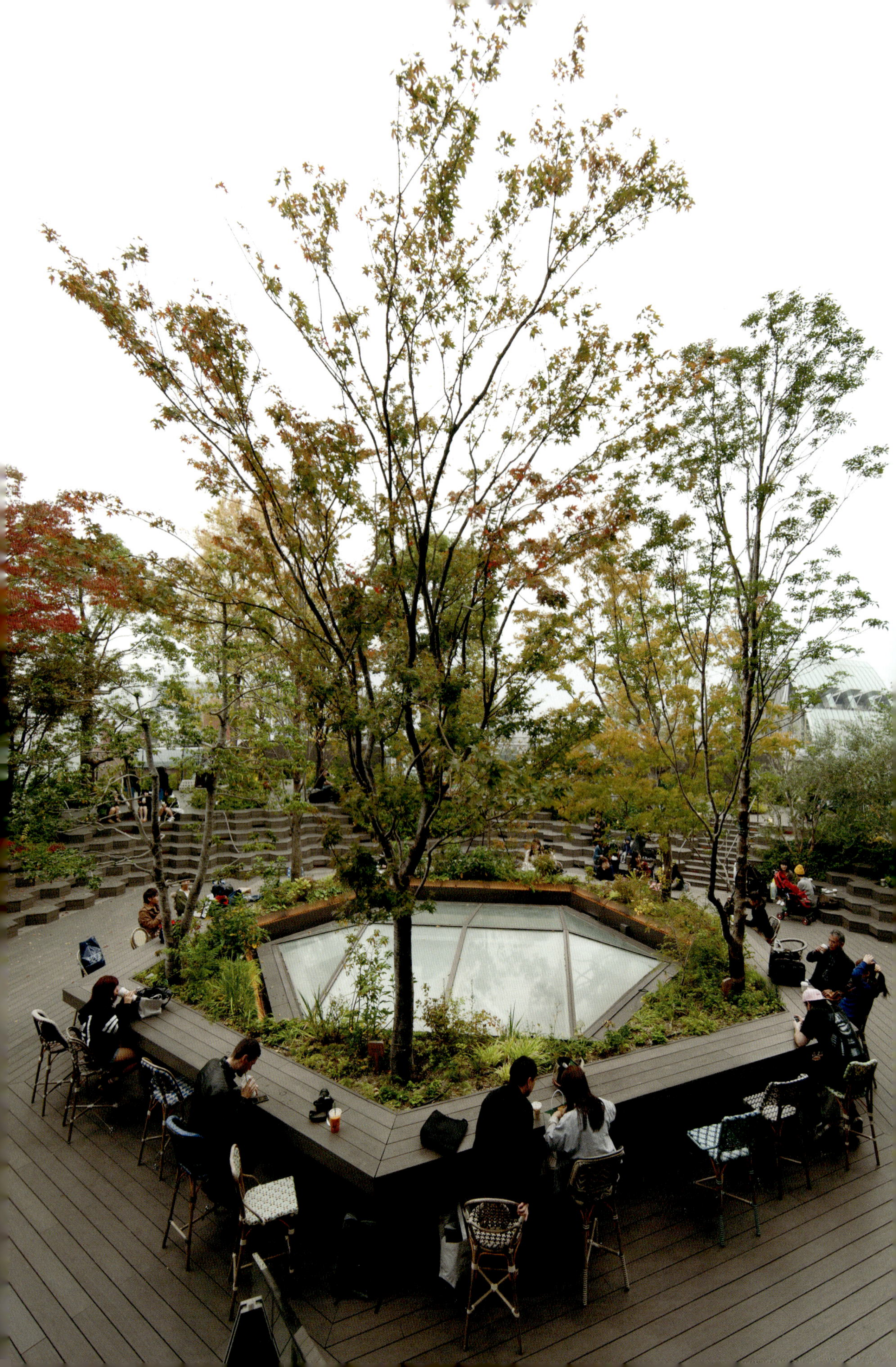

Cutting-edge architecture stands in a row and the leaves of roadside zelkova provide filtered sunlight. The designers consider the unified environment of this architectural space, combining with the space under the trees creates the quintessential "Omotesando in Tokyo" atmosphere. An environmental system which fully takes in these characteristics was proposed. In this proposal, buildings integrated with trees would float in the sky above the townscape of Omotesando, providing a succession of spaces with dappled sunlight filtering through the leaves of trees that gradually goes up to the upper parts of buildings. By adopting this proposal, birds, butterflies and the like can easily fly up to a rooftop forest and a rooftop garden is provided for people too. From a commercial perspective, the proposal produces a shower effect as customers flow from the top floor to lower floors of the building.

The initial problem was to attract customers to the 3rd or higher floors. To shorten the time of going up and down as much as possible, escalators became a space where you can experience "being in a kaleidoscope". The fashions of people on the elevators get reflected to so that they become part of a kaleidoscope and sense that they are visiting a cutting-edge town. Mirrors exaggerate the widths of the elevators make the people seem more numerous. People are tempted to come in, the same way that a queue in front of a shop tempts people to join. This building has a story: as you go up the elevator, a large atrium appears. Sunlight filters through the leaves of trees from the rooftop, and a shopping experience suitable for Omotesando begins.

Trees and terraces were placed in the area next to the buildings and the rooftop which gets the most sunshine is visible anywhere while crossing below. As a result, the inside of the building is hollowed out to allow for the leaves and the rooftop public space was raised by a height of root clumps. This inconvenience actually brings people and trees together.

The rooftop is a bowl-shaped public space by leveling off level differences by root clump with polygon steps. This open space provides ambiguous forms which are used as stairs, chairs, and tables, and this allows the users to start a physical conversation with the architecture. Visitors spontaneously turn their bodies toward the bottom of the bowl-like space and sit down as they look down to the center. People fit into small countless internal corners under dappled sunlight filtering through the leaves of trees and share the same view. Any tension associated with facing the people next is released and people feel comfortable sharing the same circle in no time. Space which is able to produce a "gentle sense of unity" of the people might become a requirement for modern commercial facilities.

録 museum
Roku museum&cafe

すべては「木ありき」。
木漏れ日が憩いの場を生み出す、
森のような美術館

Everything processed "based on the trees' existence".
Relaxing place produced by dappled sunlight filtering
through the leaves of trees at forest-like-museum

Design
中村拓志 & NAP 建築設計事務所
Hiroshi Nakamura & NAP Co., Ltd.

DATA
Site: Tochigi, Japan
Principal use: Museum, cafe
Total floor area: 106.91㎡
Completion: 9/2010
Structure: Timber structure
Structure Design: Yasushi Moribe
Contractor: Maruyama Industry Co., Ltd.

ここは栃木県の私設美術館である。交通量の多い道路沿いの敷地は元駐車場で、周囲に緑があるわけでもなかった。ここに「地域の人々が気軽に訪れるサロンのような場を建てたい」という施主の望みに応えるべく、設計者が打ち出したプランは、「まずは木をたくさん植えること」であった。緑が、人の自然な集まりを誘発させるように。その根幹にあったのは、建築設計というよりむしろ、森・公園をつくり出す感覚だったという。

設計にあたっては地元の造園業者と共に現地周辺の山の木を見て廻り、設計前にあらかじめどの樹木をどの角度に植えるかが決められた。木々は互いの枝が重ならないよう適度に離し、敷地上空がまんべんなく枝葉で覆われるよう計画し、木の枝ぶりを三次元で測量してコンピューター上にモデリングして再現。そのデータを用いて、木々の隙間を縫うように展開する建物を設計した。つまり、「まずは木ありき」ですべてが進められたのである。こうして、市街地の喧噪から隔離された、オアシスのような都市の森が誕生した。

内部空間は、身体と建築と樹木が寄り添うような親密な空間がたくさんできている。入口は上空の枝を避けるため、思わず首を縮めてしまうような天井高である。茶室のにじり口が、おじぎに似たふるまいを誘発することで人を謙虚な気持ちにさせるように、来訪者が頭を垂れることで素の自分に戻り、作品とじっくり向き合えるように、そんな願いも込められている。

枝を避けて歩きながら、日差しを避けて木陰で休む。森の中を散策するのと同じ感覚がよみがえる場である。

The Roku museum is a private museum in Tochigi Prefecture. The site of the museum used to be a parking space along a street with heavy traffic and there was no greenery around. A client wanted to build "a highly interactive place where local people can visit freely", and to achieve this, a designer planned to "plant a lot of trees first". The basic idea behind his plan was that he wanted to produce something like a forest or a park that naturally invite people to gather, rather than architectural design. In the design, the designer looked at the trees of the nearby mountains with a local landscape gardener, and he decided which trees would be planted in which directions in advance before the design process. The trees had to have an appropriate distance between them to avoid branches overlapping, with the sky above the site to have an even cover of branches and leaves. The shapes of the trees were measured in three dimensions and represented in computer modeling. The building was designed to develop as a thread between the trees with this data. In other words, everything was processed with "based on the trees' existence". In this way, an urban forest like an oasis isolated from clamor of an urban area was established.

The interior is comprised of many intimate spaces where our bodies, architecture and trees co-exist closely. To avoid the branches above, the ceiling is at a height that makes us feel that we have to duck our heads. Similar to the small door of a tea ceremony room which prompts us to make an action like a bow with modesty, at the museum visitors revert to themselves by bowing their head and deeply contemplating to art works. This is what the museum wishes upon us.

People roam among branches and take a rest in the shade of trees. In this place, one remembers the natural sense of rambling in a forest.

東京国際空港第 2 旅客ターミナルビル
Tokyo International Airport Terminal 2, "UPPER DECK TOKYO"

椅子という最小単位のケンチク群によって
巨大空間が「人間の手」に取り戻された

The huge space is back in "human hands" again, by utilizing a group of "architecture" or chairs, which are minimum architectural units.

Design
中村拓志& NAP 建築設計事務所
Hiroshi Nakamura & NAP Co., Ltd.

DATA
Site: Ota-ku, Tokyo, Japan
Principal use: Airport
Site area: approx. 159,000㎡
Completion: 10/2010
Structure: S structure
Contractor: Consortium of MHS, Taisei Corporation

公共交通施設にありがちな、ヒューマンスケールを逸脱した巨大空間をいかに人間に近しくさせるか。それが、東京国際空港（羽田空港）第2旅客ターミナル増築棟の内部空間デザインを監修した中村拓志＆NAP建築設計事務所が抱えた最初の課題であったという。

　まず2階出発ホールには、正方形の縁台のようなソファが置かれた。上部に小さな段差があり、客はそこを背もたれにしたり、荷物を置いたり、胡坐をかいたりと自由に使う。中でも設計側として想定外の喜びだったのは、四方に座った大人たちの背中の内側で子どもが遊んでいたことだという。背中のバリアに守られ安心して遊ぶ子どもたち。取材時にもソファの上で飛び跳ねる子が何人もいた。その光景からは、「君たちは、ここで自由にふるまっていいんだよ」、そんな暗黙のサインを子どもが無意識のうちに受信している様子を読み取ることができた。優しき空間提供者の声なき声が、確実に利用者に届いているのだ。

　出発ホール上階の飲食ゾーンの客席には250種の異なる椅子が配置され、ここでも客の自由なふるまいが引き出される（66・67頁写真）。一般にこのような場には大量生産の同型の椅子が並べられ、「この場にいた記憶」を呼び起こす際の妨げにすらなりかねない。だが、個性ある椅子の力が加われば、過去に座った椅子の記憶が積み重なり、場所への愛着も生まれやすい。商業的にも椅子自体が賑わいを作り出し、閑散時の寂寥感を払拭している。

　「空港」は他の公共交通施設と同様、多くの見知らぬ人々のすれ違いが繰り返されることを余儀なくされる場である。そこに「個性的な椅子」という異物が投入されることで、人々のふるまいに変化が生じる。隣に訪れたグループの、誰がどの椅子を選ぶか。人物観察の楽しみが生まれるのだ。「あの柔らかいソファを選んだあの人、意外と疲れているのかな」。ふくらむ想像は、見知らぬ他者も身近な人に感じさせる。切断されていた人間の関係性を再び取り戻す、そんな次代の公共空間の役割がここに提示されている。

We often see huge spaces that deviate from human scale in public transportation facilities in general. But, how do we make such spaces more human-friendly? It was this primary challenge that Hiroshi Nakamura & NAP faced when designing the interior space of an extended building of the Tokyo International Airport (Haneda Airport) Terminal 2.

Firstly, square sofas similar to "endai", or a traditional Japanese bench for exterior use, were placed in the departure hall on the 2nd Floor. These sofas have small steps on the upper part and customers use them as they like: as a backrest, a place to put luggage on or sitting cross-legged. It was beyond the designer's expectation to see children playing on the sofas, surrounded by their parents' backs. The children feel secure behind the "barrier" of parents' backs. We also saw a lot of children jumping around on the sofas while researching there. We could tell that children were unconsciously receiving an implicit message, "here you can do as you please". Without a voice, this gentle space still speaks to its users.

250 different chairs were placed in a restaurant zone on the upper floor of the departure hall and this also gave freedom to the behavior of users. Normally in this kind of space the same type of mass produced chairs would be placed everywhere but this may fail to leave any kind of impression and prevents people from having memories of their time spent there. By adding many unique chairs, people can remember which chair they sat in before which gives them a sense of attachment to the place. These chairs themselves create a lively commercial atmosphere and eliminate the desolation that comes during off-peak times.

In an "airport", similar to all public transport facilities, people are forced into a situation where they repeatedly pass by many strangers. But people's behavior changes with the addition of foreign objects, "unique chairs". From the group next to you, you might guess who chooses which chair. You enjoy watching people. "That person may be quite tired because he chose a soft chair". By expanding your imagination, you feel that a stranger becomes familiar. The role of the next generation public space is indicated here: mending the broken human relationship.

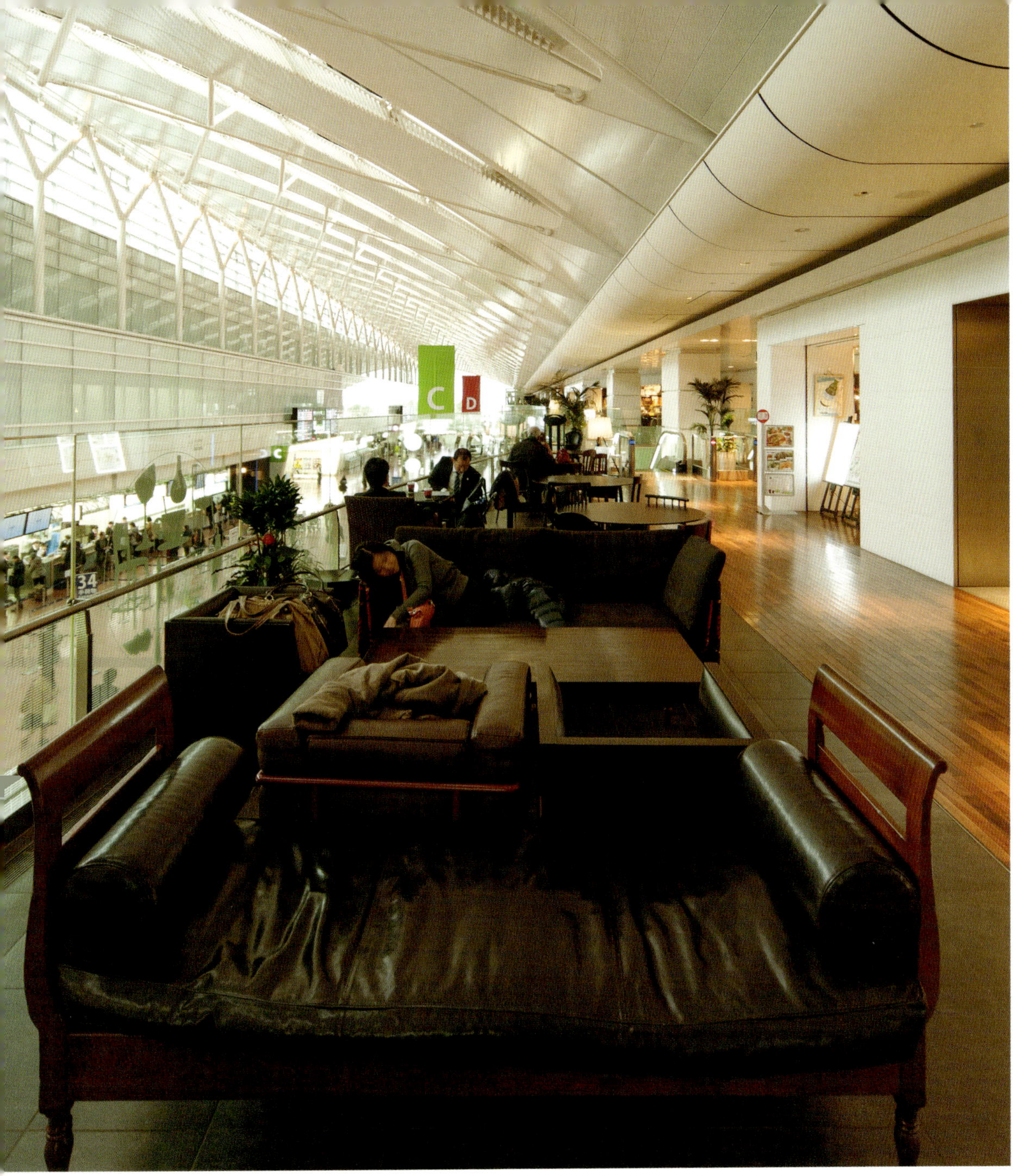
C
D
34

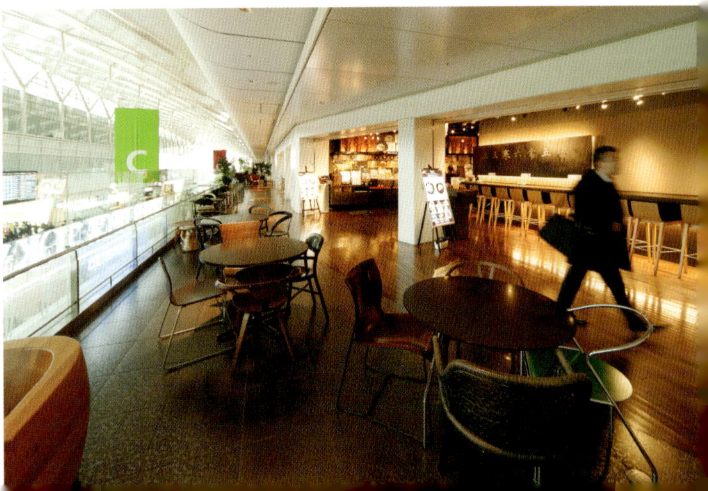
C

りくカフェ
Rikucafe

被災地の「まちのリビング」は
さまざまな人たちの思いと絆によって成長し、
地域に活力を与え続ける
*"Living room of a town" in a disaster area grows
by feelings that bind various people and energizes the area*

Design
首都大学東京　　Tokyo Metropolitan University
東京大学　　The University of Tokyo
成瀬・猪熊建築設計事務所　　Naruse-Inokuma Architects

DATA
Site: Rikuzentakata city, Iwate, Japan
Principal use: Shared space, cafe
Total floor area: 70.87㎡
Completion: 10/2014
Structure: Timber structure
Contractor: Yoshida Kensetsu

　東日本大震災の津波で平野部の主要な建物をほとんど失った陸前高田の街。震災の被害はあまりにも大きく、ダメージも癒えぬまま仮設住宅で生活する大勢の被災者がおり、人口流出も加速していた。また仮設住宅地の集会場は、利用方法の自由度が低く、部外者には使いづらい状況にあったという。

　「地域の人々が誇りに思え、力が湧き出る場をつくることで、この地は新しく生まれ変わるという希望をつくり出したい」。その思いから、行政の復興計画を待たず、高台の地に病院や食料品店などを集めた生活拠点をつくろうという住民主導の活動が生まれる。そしてその場に「誰もが集えるカフェを」という計画も浮上、そこに都市工学専門家や建築家が加わって、わずか35㎡の三角屋根の小屋がつくられた。これが前身の「りくカフェ」(仮設)である。

　仮設ゆえ予算は限られ十分な設備はなかった。

しかしその建設は、さまざまなボランティアや企業からの資材提供で支えられた。そして多くの人の想いと、地元の運営スタッフの粘り強い努力と復興への想いが実り、2014年秋から、より環境を整えたこの本設カフェがオープンした。

施設のコンセプトは、誰もが楽しく集える場、市内外を結ぶかけ橋の場、健康と生きがいづくりの場。ここでは地元住民らが煎れるコーヒーが楽しめるだけでなく、カロリーを表示した「健康ランチ」を提供し、地元の人々の生活の一部となっている。コンサートなどのイベントなども多く行われ、さまざまな使い方がなされている。地元の名産品も販売、鮮魚店の移動販売もあり、Wi-Fiも完備、仕事をする人もいるという。空間的には、勾配屋根が作る高い天井と木組みの梁が美しい、滞在を促す心地よい場が演出されている。

特筆すべきは、この空間にカフェ、販売店としての機能を導入したことで、ここが地域の集会所としてだけではなく、訪れた外部の者にも広く開かれた場になされた点だろう。また、2015年度からは、介護事業も受託予定である。ここは「都会の人と被災者との接点」であり、運営を任された地域住民に活力を与える場でもある。「これからも、地域の先進的な事例になるよう、サポートしてゆきたい」とは設計者の言葉である。

The town of Rikuzen Takada lost most of its major buildings in a plain area due to a tsunami caused by the Great East Japan Earthquake. Damage from the earthquake was too overwhelming and a lot of people lived in temporary housing, unable to recover from that traumatic event; depopulation in the town accelerated. An assembly hall within the temporary housing area did not have much flexibility for its use and it was not easy for outsiders to use the hall.

"We want to engender hope for this area to make a fresh start by creating a place which people can feel proud of and which empowers them." With this thought, residents started to take the initiative of creating a community base with a hospital, grocery stores and the like on a hill, even before any government reconstruction plan took place. Then, the plan of establishing "a cafe for everyone" at this base arose. An urban engineering specialist and an architect took part of the plan and a delta roof cabin with an area of only 35m^2 was built. This was the former "Rikucafe" (temporary construction).

The former Rikucafe had a limited budget and not enough equipment because it was a temporary construction. However, the construction was supported by various volunteers and companies offering some materials. Then, the feelings of many people together with the persistent effort and desire for reconstruction of the local administration staff resulted in a new cafe being commissioned. The new cafe would provide a better environment. It was constructed utilizing the existing temporary construction, and opened in autumn, 2014.

The facility has 3 concepts: a place where anyone can happily visit, a place that bridges the inside of the city to the outside, and a place dedicated to health and a sense of purpose in life. We can enjoy a cup of coffee that local residents make at this cafe and it provides a "healthy lunch" menu with its calories displayed. The cafe is a part of local people's lives. There are various ways to use this space. A lot of events, such as concerts, have been held here. Local specialties are sold and a mobile fish shop comes here too. You can use the Wi-Fi at the cafe and some people do their work here. The space that it provides is truly comfortable, prompting people to stay with its beautiful high ceilings under a pitched roof and exposed wooden frames.

It is especially worth mentioning that this place became not only an assembly hall for the local community but also a place wide open to visitors from outside. This was achieved with the introduction of the cafe and shop functions to the space. Also, a caretaker business is planning to be entrusted to manage the cafe as of the 2015 financial year. Rikucafe is a "point of contact where the people of the city and earthquake victims" and also a place which energizes the local administration staff. "I want to keep supporting the cafe so that it continues to be an advanced example of the community" are the words from the designer.

KOIL（柏の葉オープンイノベーションラボ）イノベーションフロア
Kashiwa-no-ha Open Innovation Lab., Innovation Floor

予想外の使われ方が生み出される、
未完成を意図した空間
Space used unexpectedly with intended incompleteness

Planning and Produce
三井不動産、ロフトワーク

Design
成瀬・猪熊建築設計事務所
Naruse-Inokuma Architects

DATA
Site: Kashiwa city, Chiba, Japan
Principal use: Co-working space, cafe, office, etc.
Floor area: 2,576㎡
Completion: 3/2014
Contractor: NOMURA Co., Ltd.

都心にほど近い千葉県柏市の柏の葉地区。かつてあったゴルフ場と米軍基地の閉鎖・移転、更に鉄道の開通などを経て、次世代型都市のモデルケースとしてさまざまな事業計画が実施されている街である。
　その一環としてオープンしたKOILは国内最大級の規模を誇るイノベーションセンターであり、その中での目玉は、約700㎡に及ぶ「KOILパーク」というコワーキングスペース[1]だ。ここは、多様な人々の交流により新たな事業・製品・サービスの生産・創造を生むプラットホームとして設けられた。そのコンセプトにふさわしい場として、従来型の無味乾燥とした、管理者にとって都合の良い均質なオフィス空間とは異なる、利用者にとって使い勝手の良いことが最優先される、多

様な性質を備えた場が必要とされていた。

そのために設計者が行ったのは、建築がそれ自体では完結しない、「未完成に留めるデザイン」である。壁は下地材の風合いを生かす仕上げとし、ホワイトボードを含むほとんどの家具にキャスターを付け、自由度を増した。更に、活発な交流を促すため、飲食・生産・くつろぎの場などを点在させ、利用者の目的に伴うアクティビティが自ずと生み出されるようにされている。さまざまな活動が同時多発的に起こる、都市のような場が実現されているのだ。

一般利用も可能な場として、3Dプリンタなどを備えたKOILファクトリー（デジタルものづくり工房）やカフェも併設された。工房は個人作家の利用も多く、カフェは隣接する大型商業施設の喧噪から逃れてきた地元民たちの隠れ家的空間にもなっている。

中央に位置するKOILスタジオ（73頁・写真下）は多様なイベントを行える場だ。「ここでミニ四駆の大会が開催された時は、そんな使い方をされるとは思ってもみなかった」と話す設計担当者の笑顔が印象的だった。決めつけないデザインから生じた「余白」が、意図を超えて活き活きと利用されるのを見届けるのは、作り手にとっても嬉しい誤算だったにちがいない。

1. 個人事業者や小規模法人がオープンスペースを共用し、独立性を保ちつつ交流を通じて互いに貢献しあう働き方やその場自体を指す。「KOILパーク」は、大企業の一部署も参加している点が従来型のコワーキングスペースと異なる大きな特色という。

The Kashiwa-no-ha area, Kashiwa in Chiba, which is near central Tokyo, was occupied by a golf course and a U.S. military base before. These facilities were closed and moved, and also a railway was opened in this area. Now, various business plans which are considered next generation model cases are being implemented for the area.

KOIL, which opened as part of the business plans, is the biggest innovation center in Japan and its main feature is a co-working space1 called "KOIL Park[1]" with an area of approximately 700m². This co-working space was created as a platform for creating new businesses, productions and services with the interaction of various people. Rather than a homogeneous office space, which is the conventional type dry space that administrators find convenient, the place required various characteristics to meet its objectives, the most important of which is usability for its inhabitants.

To create this kind place, the designer created "an intentionally incomplete design", which means that architecture is not completed by itself. A finishing of the wall was to make good use of a texture of foundation materials, and almost all furniture, including whiteboards, were equipped with casters to increase mobility and flexibility. Moreover, in order to encourage interaction between people, the place is dotted with restaurants, production studios, places for relaxation and the like to encourage to users to engage in activities naturally. The place then takes on the characteristics of a city where a multitude of activities occur simultaneously.

KOIL Factory [Digital Monozukuri (manu-facturing) Atelier] is equipped with 3D printers and the like for general use and a cafe were also established. A lot of individual artists use the atelier and the cafe is a hideaway for local people to get away from the clamor of an adjacent large commercial facility.

KOIL Studio, located in the center of the building (see the photo on the right), is a place where various events can be held. I was impressed when a designer said with a smile, "I would have never have guessed that a mini four-wheel-drive tournament would be held here". When the "blank space" that arises from a fixed upon design is used creatively beyond its intention, it must be a wonderful miscalculation for the maker.

1.way of working or a place which and where a private business or a small company share an open space and contribute each other through interaction while keeping their independence. A noticeable distinctive feature of "KOIL Park", which is different from a conventional co-working space, is that departments of large companies have also joined this space.

姫路駅北駅前広場
Himeji station North square

世界遺産のある街にふさわしい
「人間のための」広場
Public space "for humankind" suitable for a World Heritage listed town

DATA
Site: Himeji-city, Hyogo, Japan
Principal use: Station square
Site area: approx. 16,000㎡
Completion: 3/2015

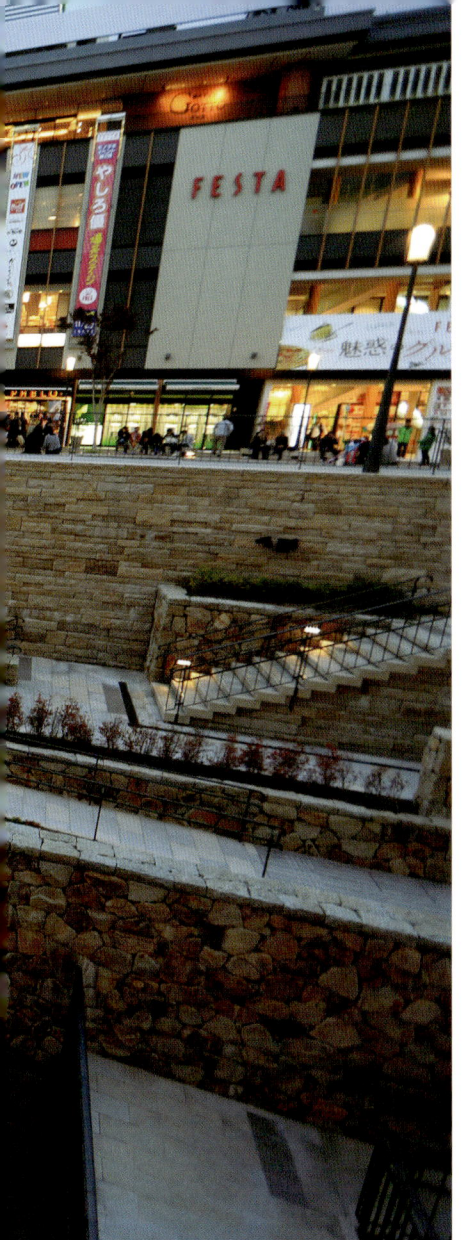

兵庫県姫路市の玄関口、駅前地区の再開発事業である。

　駅からまっすぐ延びる通りの先に、都市のシンボルである世界遺産・姫路城が位置するという日本では珍しい通景を有する駅前広場。この都市の主役は市民や観光客などの「人」であり、ここを歩行者のための空間にすることが最大のコンセプトとされた。デザインは落ち着きのある現代和風で統一され、錆御影石、地産の杉材など自然素材を多用。2階眺望デッキ、通称キャッスルビューからの城の眺めは圧巻の一言。「世界遺産のある街にふさわしい駅前広場を」という強い意思の集約から生まれた場である（43、155頁に詳細記事）。

Himeji station North square was created as a redevelopment project for the station front, which is a gateway to Himeji City in Hyogo.

　This station square has an unusual and exceptional vista in Japan, with the Himeji Castle, symbol of the city, located at the other end of a straight street from the station. "People" including citizens of the city and tourists play leading roles in the city and the most important concept was to make the space pedestrian friendly. The square is designed with a unified calm modern Japanese style where natural materials, such as yellowwish granite and local cedar wood, are frequently used. A view of the castle from a viewing deck, also known as the castle view, is simply overwhelming. This place came into being with the intensive aim of creating "a station square suitable for a World Heritage listed town" (in-depth story in p. 43 and p. 155).

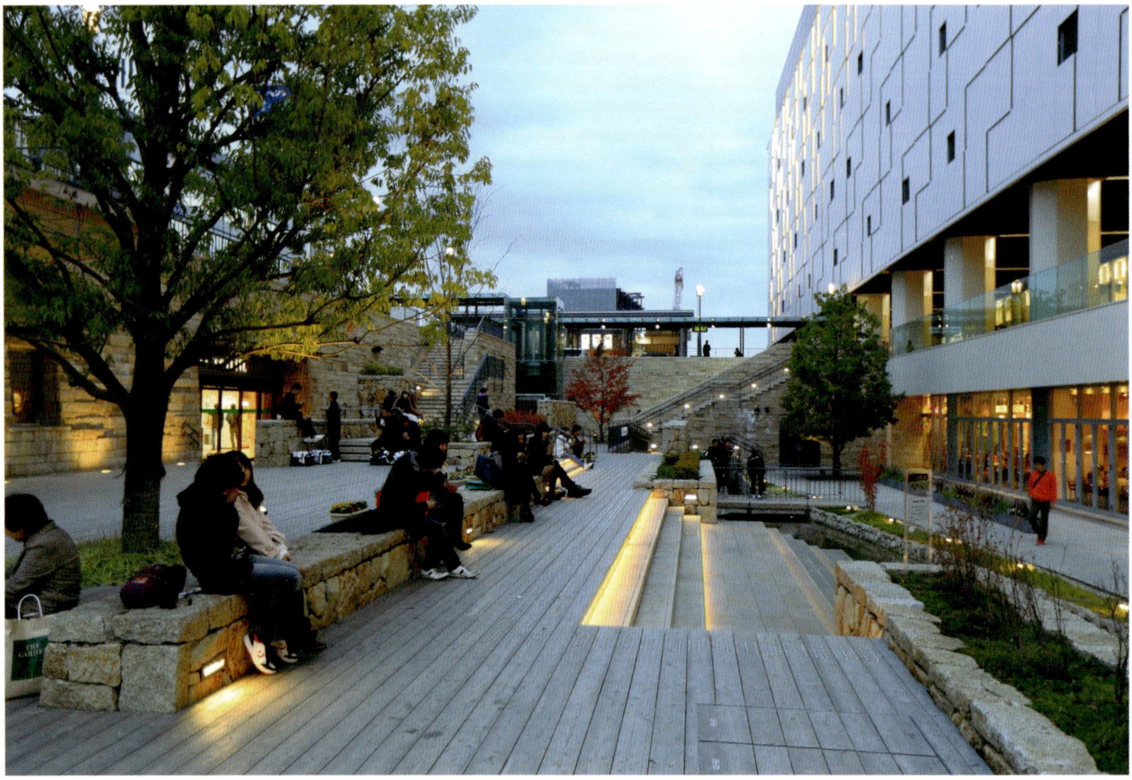

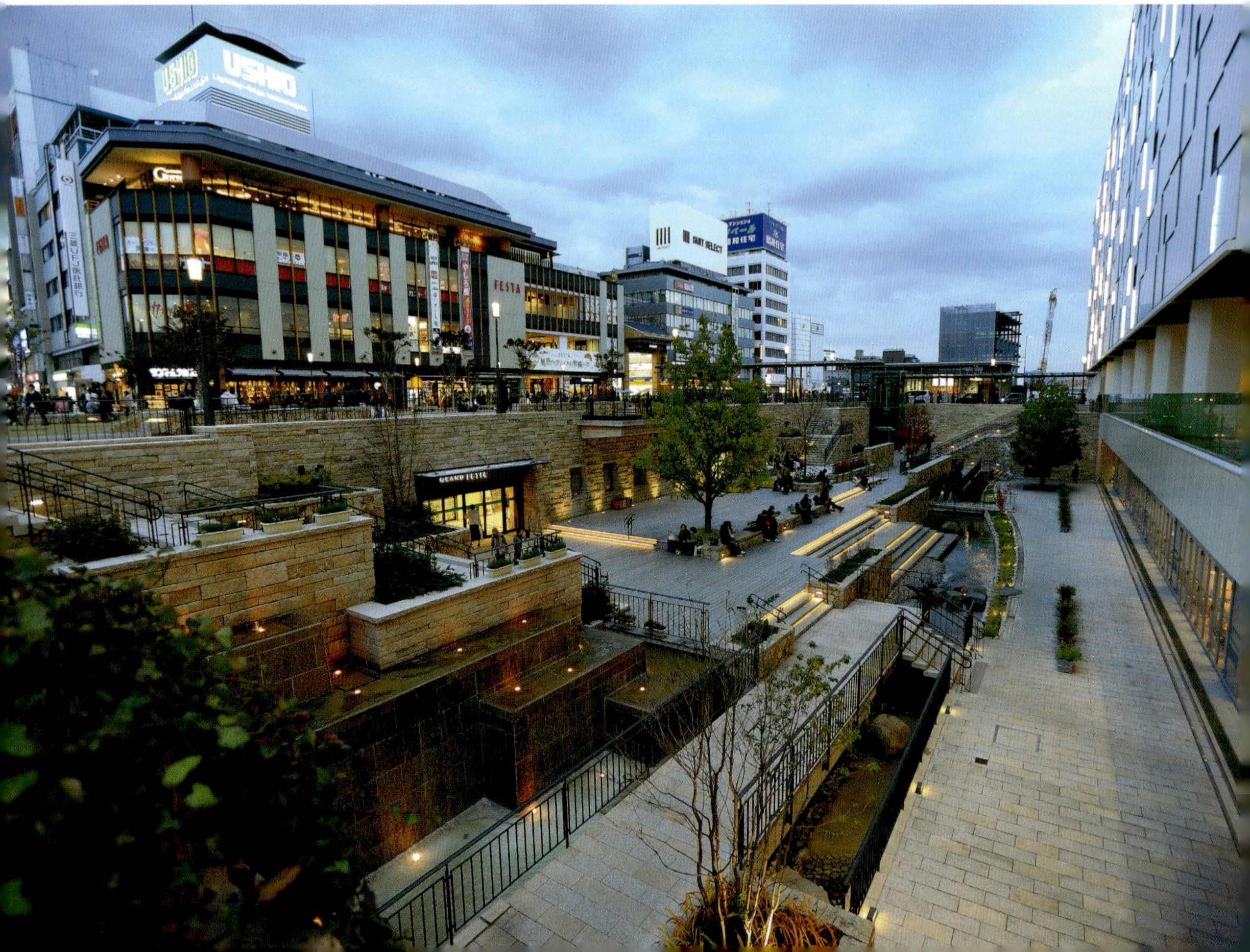
USHIO
FESTA

77

SHIBAURA HOUSE

透明感溢れる「ビル型公園」は
運河の街の新たな憩いの場

New relaxation "building type park" with full of a sense of translucence in a canal town

Design
妹島和世建築設計事務所
Kazuyo Sejima

DATA
Site: Minato-ku, Tokyo, Japan
Principal use: Office, Community space
Total floor area: 950.61㎡
Completion: 7/2011
Structure: S structure
Contractor: SHIMIZU CORPORATION

　東京都港区芝浦は、運河の合間を縫ってオフィスビルと高層マンションが建ち並ぶ新興の地である。最寄り駅の利用者は一日約15万人、その大半を勤め人が占め、高層マンションには子育て世代の移住者も多いものの、公園などの公共施設は少なく、コミュニティが成り立ちやすいとは言い難い環境にあった。
　この地にオープンしたSHIBAURA HOUSEは、「よく間違われる」そうだが公共の施設ではない。老舗の広告制作会社の自社ビルである。社屋建て替えの際、「せっかくなら地域の人々に使えるものを」と着想し、設計を妹島和世に依頼。2層吹き抜けが駆使された5階建てのビルが完成した。5階は全面ガラス張りの開放感溢れる空間で、専用テラスやプロジェクターも備え多様な催しに対応できるレンタルスペースである。4階

は自社オフィス、3階は中庭付きのレンタルフロア。そして2階と1階が、誰もが無料で利用できるフリースペースである（イベント利用時など一部時間を除く）。

　2階はカウンターと小さなテーブルが配置され、グラウンドレベルから隔離されることで、やや奥まった都市の隠れ家、いわば「避難地」のような空間が構成され、空間の居心地の良さに引き

Shibaura, Minato-ku in Tokyo is a new area where office buildings and high-rise apartments stand in a row surrounded by the network of canals. Approximately 150,000 people use the nearest station everyday and most of them are office workers. Although a lot of child rearing couples have moved to this area, its environment is not suitable for a local community to establish itself easily because it lacks public facilities such as parks.

　SHIBAURA HOUSE, which opened in this area, is often "misjudged" as a public facility, but the building is actually the in-house building of a long-established advertising production company. The company had an idea of how its building would be used when it was rebuilt; "since it is going to be rebuilt, the building should be used by local people". Kazuyo Sejima designed it with this idea and a 5 story building with 2 layers of open ceiling space was completed. The 5th floor is a glazed space filled with the sense of openness. It is a rental space with an

5th Floor

3rd Floor

exclusive terrace and a projector and the space can be used for various events. An in-house office is located in the 4th floor and the 3rd floor is a rental space with a courtyard. The 1st and 2nd floors are free spaces which anyone can use free of charge (except when it is used for an event, etc.).

Counters and small tables were located on the 2nd floor and a space like a secluded urban hideaway or a safe haven, so to speak, was created by isolating it from the ground level. A lot of women gather there as they are drawn to the comfortableness of the space. Various people including office workers, housewives with prams taking shelter from the rain and old ladies in the neighborhood gather in the bright space on the 1st floor with the area of 100m^2 with a ceiling height of over 5m. Various events, such as craft classes for children and English conversation classes, are held daily at one corner on the 1st floor.

Adjacent spaces, such as upper and lower floors, or inside and outside, are loosely connected so that people can feel a sense of togetherness throughout the building, and such spatial continuity makes this space very attractive. The atmosphere here reminds us of a scene in an old alleyway where we would see, "old ladies warning children who are running around".

寄せられるように女性たちが多く集まっていた。広さ100㎡、天井高5mを超す明るい1階も、会社員やベビーカーを押す雨宿りの主婦、そして近所のおばあちゃんまで、さまざまな人々が集まってくる。その一角では子ども工作教室や英会話教室など、日替わりで多様なイベントが行われている。

ある階から別の階、そして外部と、どこにいても隣接する場を感じることのできる緩やかな連動性が、空間に大きな魅力を与えている。「走り回る子どもをおばあちゃんが注意する」、かつての路地空間で見られたような光景が、この場で再現されているのだ。

2nd Floor

木屋旅館
KIYA RYOKAN

引き算の手法でポテンシャルを引き出す
街のシンボルに生まれ変わった古建築

Exploiting potential by a method of subtraction
Historical architecture reborn as a town symbol

Design
永山祐子建築設計
YUKO NAGAYAMA&ASSOCIATES

DATA
Site: Uwajima city, Ehime, Japan
Principal use: Japanese-style hotel
Site area: 559㎡
Completion: 11/2011
Structure: Timber structure
Contractor: Miyata Construction

闘牛などの独自文化が色濃く残る南海の地方都市・宇和島の地にオープンした滞在型の観光名所、木屋旅館。ここは滞在者が旅館1棟を2～10人の1組限定で借り切り、自由に建物を使えるというスタイルを採用している。明治時代の旅籠の面影が残る木造2階建ての家屋のリノベーションを担当した建築家の永山祐子氏は、「そこには100年の歳月をかけ熟成された代え難い空間が存在していた。そこに何かを足すのではなく、あえて引き算をすることで、既存の物語の新たな一面が発見され、新たな旅館の価値になるのでは」と設計コンセプトを明かす。

　まず、水平方向のみしか視線が抜けなかった既存空間に対し、垂直方向に空間を引く試みがなされる。2階の床を広範囲で抜き透明アクリル床を配置。さらにその上部の天井も抜き、古い屋根の架構を露出させた。最大約8mの高さの断面が創り出す新しい視点。吹き抜けを通して1階にも光が差す。夜、ファサードからは障子を通して行灯のような光が街に漏れ出す。その佇まいが、宇和島という街全体へのアクションになればという試みだ。なおオープン後、本旅館を舞台にアートイベント「AT ART UWAJIMA2013」が行われるなど、宿泊以外にも利用されている（161頁に詳細記事）。

The Kiya Ryokan provides Japanese style accommodation, popular for tourists. It was opened in the Uwajima provisional city located in the Nankai area which has a unique culture that remains strong, including a bullfight. This accommodation adopts a policy which allows only one group of people of 2-10 people to stay in the building at a time and they can use the building to their liking. The architect, Yuko Nagayama, was in charge of renovating the 2 story wooden house which still has the look of an inn from the Meiji period. She revealed the design concept of this accommodation: "developed, irreplaceable space existed here after 100 years. I thought, instead of adding, we would be able to find a new face of an existing story by purposefully subtracting, and it would create new value out of the accommodation."

　The first step was to create a vertical space in the building where only horizontal space existed. To achieve this, a wide floor area in the 2nd floor was removed and a transparent acrylic floor was placed there. A ceiling above the transparent floor area was also removed and the old roof beams were exposed. A section with a height of approximately 8m creates a new viewpoint; a vertical space atrium. The sun shines through to the 1st floor through the atrium. The lantern-like light coming through a shoji (paper screen) leaks out from the facade to the town at night. The purpose of this appearance is to have an effect on the whole city, Uwajima. After the open of the hotel, the place is used for other uses than a hotel, such as an art event venue "AT ART UWAJIMA2013" (in-depth story in p. 161).

ルイーズ・ブルジョワ《ママン》2002（1999）
Louise Bourgeois, "MAMAN", 2002 (1999)

　ビジネスマンから旅行客まで、世界中から人を集める東京・六本木ヒルズ。ここを都心の文化の中心地とするアイディアの一環として、敷地内の各所に世界中の芸術家の作品が設置されるプロジェクトが進行。玄関口のパブリックエリアである66プラザに、『ママン』と題された高さ10メートルの巨大なクモのオブジェが設置された。同作は彫刻家のルイーズ・ブルジョア（1911〜2010）の作品で、その敬愛する聡明な母親がモチーフという。一見グロテスクだが大きな包容力をも感じさせるその姿は多くの人々の記憶に留まり、格好の待ち合わせ場所として、都心の新たな景観を形づくっている。

六本木ヒルズ 66プラザ
Roppongi Hills 66 Plaza

訪れた者の記憶に留まり続ける
都心のユニークな待ち合わせ場所

*A unique meeting point in the city center
that remains in visitors memories*

DATA
Site: Minato-ku, Tokyo, Japan
Principal use: Public square
Site area: approx. 5,150㎡
Completion: 2003

Roppongi Hills in Tokyo attracts people from all over the world, including business people and travelers. As part of an idea to make this place a hub for urban cultures, a project to display works of international artists in various places of this space is under way. A public artwork of a ten meter high gigantic spider titled "MAMAN" was installed in the public area of the gateway, Roku-Roku Plaza. This work is by the sculptor Louise Bourgeois (1911-2010), and the motif of the work is her sagacious mother that she adored. Despite its seemingly grotesque look, the sculpture with an accepting feel remains in people's minds, and it creates a new scene in the city as a meeting point.

仏生山温泉（仏生山まちぐるみ旅館）
Busshozan-onsen (Busshozan-machigurumi-hotel)

旅館を軸に、街を更新する
めざすのは居心地良い場の提供

Renew the town centered around the hotel
Aim to provide a comfortable place

Design
設計事務所岡昇平
Shohei Oka

Site: Takamatsu city, Kagawa, Japan
Principal use: Spa facility
Total floor area: 622.98㎡
Completion: 2005
Structure: S structure
Contractor: Taniguchi Architects & Devisers

香川県高松市の郊外住宅地に所在する温泉施設である。

　ここは江戸時代、高松藩主の菩提寺の門前町として栄えたが、近年は幹線道路沿いに大型店舗が建ち並ぶ「車のための街」と化していた。この地で飲食業を営む主が温泉を掘り当てたことが、「仏生山温泉」誕生の契機となる。

　温泉施設の設計を担当し、運営も任されたのはその主の子息で、「みかんぐみ」に勤めたのち地元で独立した建築家・岡昇平氏である。建物は一目では浴場とわからない洒脱さで、フロント・通路・売店・休憩所が一体となった細長い空間で構成され、祭りやフリーマーケットが開催されるなど、地域の多様なコミュニケーションの場ともなっている。大きくとられた開口部は開放感を与え、浴場も露天風呂と脱衣所など全体がひとつながりの空間として捉えられ、自由度の高い利用が可能だ。岡氏は「いちばん気を配ったのは、温泉施設としての居心地の良さ」という。その思いは、売店で古本を扱い、浴場内への持ち込みも可能にしたユニークなシステムを導入したことにも

Busshozan-onsen is a hot spring facility in a residential suburb of Takamatsu City, Kagawa.

　This area flourished as a Monzenmachi (temple town) of the Takamatsu feudal lord's Bodai-ji Temple in the Edo period, but has, in recent years, became "a town for cars", where large shops line up along main roads. The defining moment for "Busshozan-onsen" was when a restaurant owner struck a hot spring source in the area.

　An architect son of the restaurant owner, Shohei Oka, who set up his office locally after working for MIKAN, designed and also manages the hot spring facility. Unlike most Japanese hot springs, its building is stylish which means that people don't initially recognize it as a public bath. The building has a long and narrow unified space for the reception, corridors, shops and rest area. This serves as a communication space for the community in a variety of ways, for things such as festivals and flea markets. A large opening gives us a sense of openness, and we can use the space freely because the whole area including the bath area, outdoor hot spring and changing room is grasped as a merged space. Oka said, "I paid the

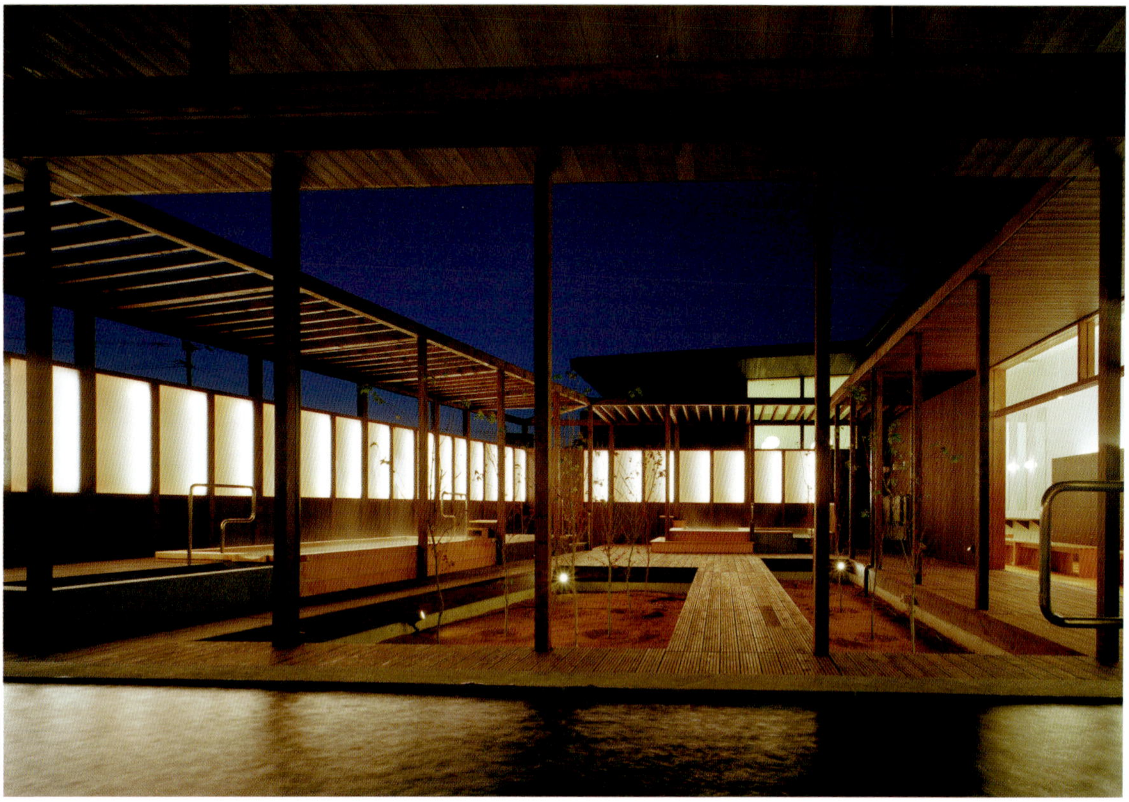

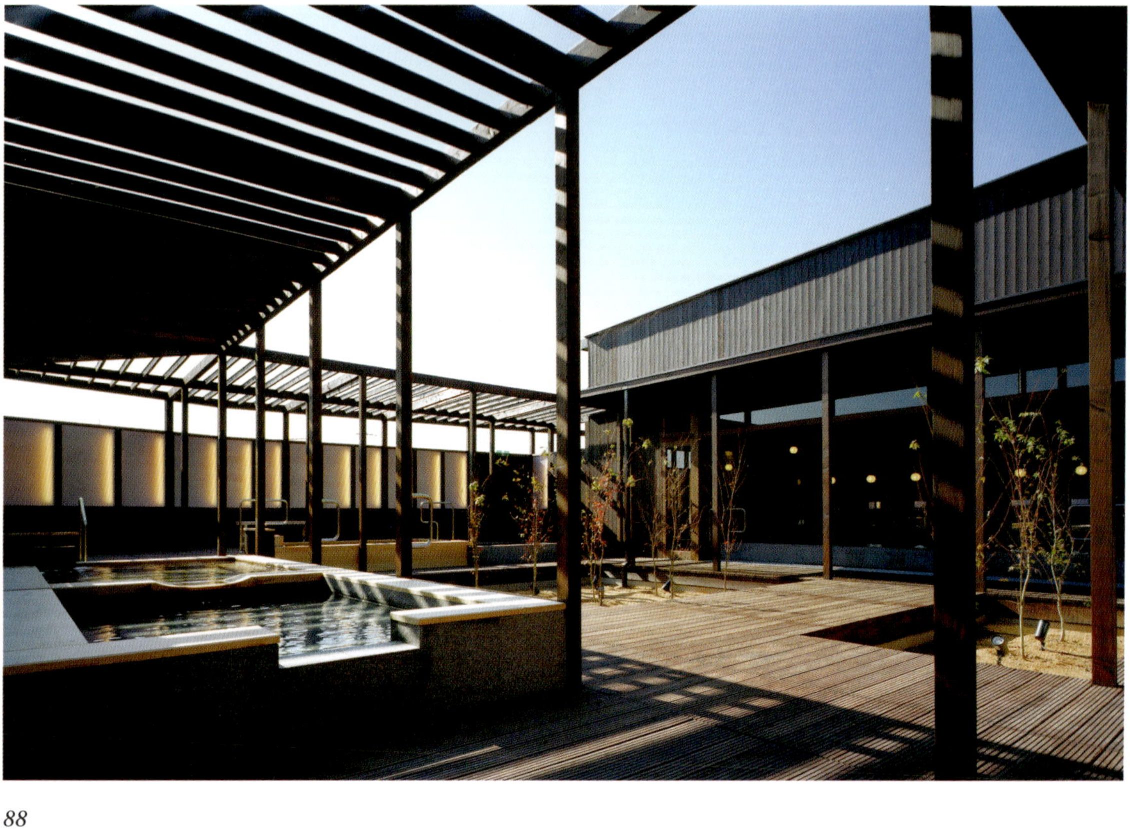

現れている。ここでは客の回転率を上げることではなく、居心地の良さを高めることが最優先されているのだ。

　ユニークな取り組みは施設内にとどまらない。温泉に宿泊施設はないが、岡氏の働きかけで、まちの離れた地に宿泊棟や食事どころ・カフェなどが新設され、また既設店舗の協力も得て、全体として旅館の機能を持つようまち自体が変わっていったのである。まち単位で来訪者をもてなそうという「まちぐるみ旅館」活動の始まりだ。実際、温泉の質や施設空間の心地よさ、そしてまちぐるみ旅館の活動が評判を呼び、県内外から多くの顧客を呼び込むことに成功している。

　ただ、活動としては「枠をつくらず、日々の暮らしの中からまちがよくなるように」とあえて組織化せず自主性に任せ、ゆるやかな発展を望んでいるという。自由度が高く誰もが参加しやすいその取り組み方は、一地方都市における街の振興活動の新たなモデルケースとしても注目を集めている。

　「ただ自分が楽しく居心地の良い場をつくりたかった」と岡氏。建築ではなく、「場」をつくるという発想こそ、今後の公共空間に最も求められる要件かもしれない。

closest attention to comfortableness as a hot spring facility". A unique system whereby customers can buy secondhand books at a shop and bring them into the bath area reflects his thoughtfulness. To increase comfortableness is more important than to increase a customer turnover rate in this hot spring facility.

The uniqueness can be found not only in the facility. This hot spring facility does not have an accommodation facility; however, accommodations, restaurants, cafes and the like were newly established in places away from the hot spring facility in the town by Oka's approach. Also, the town itself had changed to have a hotel function as a whole with the cooperation of existing shops. The "machigurumi-hotel" (hotel as a whole town) activity which entertains guests as a town unit started. In fact, the hot spring facility became very popular because of the quality of hot spring, comfortableness of the facility space and machigurumi-hotel activity and to attract a lot of customers from outside the prefecture has succeeded.

The activity is hoped to develop slowly without organized and with respecting independences: "to improve the town from everyday life without making frameworks". Anyone can easily join the activity freely and this method is getting attention as a new model case of a town promotional activity in one provisional city.

Oka said "I just wanted to create a comfortable place with fun". The most important requirement for a future public space may be a way of thinking to create a "place" but not architecture.

代官山ヒルサイドテラス
HILLSIDE TERRACE

都市の変容と共に歩み続ける「アーバンヴィレッジ」
An "urban village" that develops together with transformation of the city

Design
槇総合計画事務所
Maki and Associates

DATA
Site: Shibuya-ku, Tokyo, Japan
Principal use: Store, Residence, Office
Completion: 1969 – 1992
Structure: RC structure
Contractor: Takenaka Corporation

　いわずと知れたこの名建築は、「代官山」というこの地の都市文化の成立に深く影響を与えてきた。敷地には古墳もあり、ゲニウス・ロキ（地霊）を尊重することは設計者と地主の共通見解であったという。性急な開発を望まない両者によって、住居と商業空間、プライバシーとパブリシティが両立する開発計画が数期に分け作成された。11棟からなる建物は建築時期により形態やファサードが異なるものの、そこに自然な統一感があるのは、すべて「人間的尺度」が基準とされているからなのだろう。歩道や階段は広々とし、まさに「人のために」つくられた場として、住民や働く者、通行人らが交じり合う、高いアクティビティを誇る移行空間となっている。

　内部に入るとその先にはギャラリーがあり窓があり、ポケットパークやショップが連なる。幾重

もの「層」と、向こう側に何かあると思わせる「連続性」により、限られた空間に奥行きが生み出される。通りを軸とし何層もの緩やかな境界を設けて空間に「奥性」を与えていく、日本の伝統的な建築手法がここに生かされている。

　F棟とG棟の間には開けた小広場がある。「ヒルサイドスクエア」と呼ばれるこの場は、計画当初から「広場」としての使用が意図されたわけではなく、マーケットやイベントなどで自然発生的に活用されることが増え、後にこの名が付けられたという。通行人もふと足を止める、都市の延長線上にある広場。一つの開発計画で完結することなく、移りゆく時代を見据え、場のもつ価値を維持し高め続ける。年月をかけ熟成を遂げた街の姿を象徴するかのような大木が、広場に心地よい木漏れ日を落とす。

This established architectural masterpiece has been providing a deep effect to the formation of the urban culture of the area, "Daikan-yama". There is even an ancient burial mound in the premises, and it was a common understanding between the designer and the land owner to pay respect to Genius Loci. Neither the designer nor the land owner wished for a speedy development, which resulted in the creation of several separate development plans to combine residential and commercial areas, and private and public lives. Although the property of 11 buildings has variations in their appearance and facade depending on the time of completion, the buildings may owe the sense of natural unity to sharing the "human scale" standard. Paths and staircases are wide, and serving as a passage, designed "just for humans", for vigorous activities where the residents, workers and passersby can interact with one another.

　As you enter, there is a gallery and windows, pocket parks and shops one after another. Several "layers" and the "continuity" suggest something is awaiting on the other side and create depth in the limited space. A traditional Japanese architectural technique, to provide the sense of "depth" to space by adding gradual border lines on the both sides of a corridor, is well observed here.

　There is a small public space between Building F and Building G. This space is now called "Hillside Square" but it was not designed as a "Hiroba" (public space) at the initial stages of planning, however, it gradually started to be used for public markets and events, which later resulted in the space acquiring its name, "Hillside Square". This public space is regarded as an extension of the city where passersby stop by. It was not completed in one developmental plan, but rather it continues to maintain and enhance values of the land in line with changing times. A magnificent tree, which can be seen as a symbol of this town that has matured over time, casts pleasant dappled sunlight filtering through its leaves.

代官山 蔦屋書店
DAIKANYAMA T-SITE

名建築が築いた街に生まれた文化の森
Culture Forest has appeared in a town with fine architecture

Design
クライン ダイサム アーキテクツ + RIA
Klein Dytham architecture + RIA

DATA
Site: Shibuya-ku, Tokyo, Japan
Principal use: Store
Total floor area: 1,387㎡（1号館）1,288㎡（2号館）1,530㎡（3号館）
Completion: 12/2011
Structure: S structure
Contractor: Kajima Corporation

「文化の森を創る、文化のインフラを創る」ことがコンセプトのこの複合書店は、一帯のランドマークであるヒルサイドテラス（90・91頁）に隣接する。ここは水戸徳川家の屋敷跡地という由緒正しい土地で、さらにさかのぼれば、この高台の地が数千年以上前からの「人類にとっての一等地」であることは、近隣に在る古墳などからも証明されている。

3棟からなる店内は書店などのほか、いたるところにくつろぎのスペースもあり、それらが渾然一体となって人々の居場所を多様に演出する。シンプルな箱状の3つの建物は間を空けて配置され、露地のような広場のような空間を作り出す。植え込みの表情は繊細で、既存の樹木も切られることなく生かされている。専門店群が散歩道で繋がるエリアはぶらつくだけでも心地よい。約4000坪の敷地を存分に活用し、好きな本や映画探しに没頭でき、飲食にくつろげる心地よい都市

生活を人々に提供している。

　ヒルサイドテラスという、街の文化を創り育んできた名建築のある地に、文化を発掘し、文化に触れられる場所がまた新たに誕生した。すべては時の流れの中の、必然の出来事なのかもしれない。

This compound bookstore, created with the concept "to make infrastructure of the culture, to make a forest of the culture", is located next to Hillside Terrace (pp. 90-91), a landmark of this area. This place is the noble land which traces its ownership back to the Mito Tokugawa family. Going further into the past, it is proven from the old burial mounds around the area, that this upland is "an excellent location for humans" and has been for several thousand years.

　There are three buildings and within them are a bookstore and relaxation spaces creating places where various people can come together in complete harmony. Three simple, box-shaped buildings with gaps in between, create space, such as, the open space or the public space. The expression of the shrubbery is delicate, and old trees exist without ever being cut. It is comfortable just to stroll around this area where specialty stores lead onto a promenade. Fully utilizing the site of approximately 13,200m^2, offers a comfortable city life for all people devoted to searching for favorite books or movies, and relaxing while eating and drinking.

　A place where culture can be excavated and experienced has appeared again recently in the area, with the famous building, Hillside Terrace, where the culture of the town has been created and kept alive. This is all a necessity in the passage of time.

マーチエキュート神田万世橋
mAAch ecute KANDA MANSEIBASHI

鉄道高架橋のアーチを利用した
人を惹きつけてやまない、新旧両存のスポット

Attractive place combining the old and the new
Utilizing the arch of a railway viaduct

Design
東日本旅客鉄道　ジェイアール東日本建築設計事務所　みかんぐみ
East Japan Railway Company, JR East Design Corporation, MIKAN

DATA
Site: Chiyoda-ku, Tokyo, Japan
Principal use: Commercial facility
Total floor area: 2592.48㎡
Completion: 8/2013
Structure: S structure, Brick structure
Contractor: TODA CORPORATION, Totetsu Kogyo Co., Ltd

神田川のほとりに立つレンガのアーチ。ここはかつて中央線のターミナル「万世橋駅」があった地だ。駅舎の設計は東京駅と同じ辰野金吾で、関東大震災などを経て次第に規模を縮小し、1936年から駅舎は鉄道博物館に建て替えられた。このアーチ高架橋下も博物館展示室の一部だったが、閉館に伴い、改修し2013年に商業施設へと生まれ変わった。

売り場と売り場がアーチのトンネルでつながる内部空間、神田川を一望できるオープンデッキ、走行中の電車を眺められるガラス張りのプラットホームなど見どころは多い。また、隣接するJR神田万世橋ビルとの空間に「サウスコリドー」と呼ばれる通りがあり、そこにはベンチが設置され[1]、自然と人が集まる憩いの場となっている。古くて新しい空間を体感できる、東京の新名所である。

mAAch ecute KANDA MANSEIBASHI, originally built in 1912, was previously a terminal of the Chuo Line, known as Manseibashi station. Kingo Tatsuno, his architect, was also famous for designing Tokyo Station. However, after the Great Kanto earthquake of 1923 the station building was gradually scaled down prior being rebuilt as the Railway Museum in 1936. Finally the museum was torn down and the arch viaduct next to it was closed until its renovation as a commercial facility in 2013.

Highlights of the new facility are numerous. They include an inner sales floor area connected by a tunnel made up with concrete and brick arches, and a new open deck from where one can rediscover the Kanda River. Last but not least, a glazed platform and a cafe, which are literally surrounded by trains in operation, top the viaduct. After enjoying this spectacular set of spaces, one can eventually reach a pedestrian street called the "South Corridor" which is adjacent to the Japan Railways (JR) Kanda Manseibashi Building. Here, benches stand[1] freely, creating a relaxing space where people gather naturally, enjoying Tokyo both old and new.

1. 常設ではない

1. Not for a permanent use

東北自動車道　羽生パーキングエリア（上り線）
The Tohoku Expressway Hanyu Parking Area (Up Line)

ハイウェイに突如現れた江戸の街並み。
話題性と象徴性、高い完成度を備えた
現代日本の特異空間

Edo townscape suddenly appearing on an expressway
Special space in contemporary Japan
with topicality, symbolism and high completeness

DATA
Site: Hanyu city, Saitama, Japan
Principal use: Service area
Total floor area: approx. 1,000㎡
Completion: 12/2013
Structure: S structure
Design: Takanori Aiba
Consulting and supervision: NOMURA Co., Ltd.
Contractor: Tanakakensetu Co.,Ltd.

高速道路に江戸の街並み。この意外な組み合わせが話題となっている。

　この地には江戸時代、江戸と東北を結ぶ街道上に位置する関所があった。地理的にも歴史的にも「江戸の入り口」だった場に作られた、江戸の世界観に触れられるテーマパーク型のパーキングエリア（以下PA）である。ここは「人気の時代小説の舞台の再現」がコンセプトでもあり、その世界観を忠実に表すため、新築した江戸時代風の建物にはエイジング加工が施され、その完成度の高さも人を引きつける一因となっている。

　若年世代には珍しく、中高年には馴染み深い。外国人にとっても、日本の伝統に気軽に触れられるだけでなく、日本の"ガラパゴス文化"を体験できる貴重なスポットとなっている。

　観覧車や遊園施設、温泉などを備えたテーマパーク型PAは、ほかにも国内各地に存在する。中でもこの羽生PAは、清潔さと利便性、食事や土産品の充実など、顧客満足を追求してきた日本のハイウェイのサービスエリアという特異空間の、一つの到達点といえるかもしれない。

An expressway with an Edo period townscape – this unexpected combination is now topical.

　The Hanyu Parking Area (up line) is in a place where checkpoints and Sekisho (barriers) existed on a street from Edo through to the Tohoku area during the Edo period. It is a themed parking area (herein after referred to as PA) where you can see the outlook from Edo and it was built where the "entrance of Edo" was, both geographically and historically. The concept of this PA is the "reproduction of a stage of a popular historical novel". A new building was constructed with Edo period styling and weathered-look finish was applied to it so that the concept was faithfully executed. The completeness of the weathered look of the building is a major factor in attracting people to it.

　This PA is unusual for the younger generation, but is familiar to the middle and old aged generations. For foreign visitors, it is a valuable place where they can experience not only Japanese traditions but also the Japanese "Galapagos culture" (i.e. unique culture not subject to outside influences).

　Other themed PAs with Ferris wheels, playground facilities, hot springs and the like exist in various places in Japan. Service areas on Japanese expressways have pursued customer satisfaction with cleanness, convenience, meals, souvenirs and the like. As an arrival point of an expressway service area, the Hanyu PA stands out exceptionally.

サンストリート亀戸
Sun Street KAMEIDO

親しみを感じることを禁じえない、
ほどよい空間スケール。
地域に欠かせない生活の中心地だからこそ、
新陳代謝のスピードは日々増し続ける。

Space of moderate scale with insuppressible familiarity
A vital center of life for the community
increases metabolism every day

Design and supervision
北山孝二郎 + K計画事務所
Kojiro Kitayama + K Architect & Associates

DATA
Site: Koto-ku, Tokyo, Japan
Principal use: Shopping center
Total floor area: 15,871 ㎡
Completion: 11/1997
Structure: S structure

東京都心部の東側、いわゆる下町地区に位置し、幹線道路沿いで車でのアクセスも良い。しかし典型的な郊外型ショッピングモールと決定的に異なるのは「徒歩で入場する人の多さ」である。駅に近い立地の良さもあるが、特筆すべきは正面エントランス以外にも各所に設けられた出入り口や階段から、敷地内（建物2階にさえ！）にダイレクトに入場できるなど、周囲との高い接続性を備えている点だろう。

元は時計会社の工場跡地で、当初の高層ビル開発計画の凍結に伴い、期間限定の暫定利用での商業施設計画に至ったという。その「暫定利用＝解体しやすい小規模開発を」というコンセプトのために、極めて容積率消化の小さい低層施設となった点が、広すぎず狭すぎず、ゆとりあるスケール感につながっているといえる。

空間デザインの特徴としては、開放的でありながらもほどよく閉鎖的な約900㎡の「マーケット広場」にまず目が向かう。入り口から広場へと続

く逆S字カーブの通路は見通しが悪く、その通路を抜け広場に出ると急激に視界が開けることで開放感を演出している。通路沿いの店は露店のように屋外にも商品を陳列するなどして個性を競い合い、ランブリング（ぶらぶら歩き）の楽しみを誘発させている。

広場では毎日のようにさまざまなイベントが開かれている。地域住民の生活に欠かせない場であることは、1,000万人に達するという年間来場者数からもみてとれる。

The Sun Street Kameido shopping centre is located in the so-called Shitamachi area, a low-lying area on the east side of central Tokyo near Tokyo Bay. It is adjacent to a main road so it is easily accessible by car, however, it is decidedly different from a typical suburban shopping mall because "many people go there by foot". One of the reasons is its good location near a station, but a more important point to note is its high connectivity with surrounding areas. For example, you can enter the premises (even the 2nd floor areas!) directly not only from a front entrance but also from other gateways and stairs.

The freezing of the first high-rise building development plan at the site of a demolished clock factory led to an alternative plan for a temporary commercial facility for the limited time only. The temporary use required a small scale development that is easily dismantled. The result is a low-rise facility with a high floor-area ratio which leads to a sense of comfortable scale, which is neither too big nor too small.

The main characteristic of the design is an "open market space" of approximately 900m^2, which is predominantly open but partially closed. The sense of openness is enhanced because of a reverse S-curve aisle which leads from the entrance the open space. This means visitors don't see the open space until it suddenly and majestically presents itself as they come around bend. Shops along the aisle put their products outside like a street stall to compete for individuality and this tempts visitors to enjoy a ramble.

Various events are held at the open space almost every day. This open space is a vital place for local residents' lives, a fact borne out by annual visitor numbers reaching 10 million.

権堂パブリックスペースOPEN
Gondo Public space OPEN

古建築を「更新」し、
その魅力を共有するサロンを創造する
"Renewal" of historical buildings.
Creating a place for interaction to share the charms

DATA
Site: Nagano city, Japan
Principal use: Share office, shop, cafe, co-working space
Site area: approx. 700㎡
Completion: 6/2012

Gondo-cho Town is located on the south side of a large temple, Zenkoji Temple, enshrined in a center of Nagano City. The town used to be the biggest downtown area of the city. Here, a 2-story residence and 2 warehouses were lined up together, which were approximately 100 years old. In the residence there was also a kimono wholesale shop. When the buildings were set for demolition, people declared them "too valuable to demolish". Soon multiple tenants signed a contract and a public space "OPEN" was created based on a rule whereby tenants would manage the buildings by cooperation.

Tenants renovated the interiors of the buildings in a variety of ways. The 1st floor of the residence is a cafe and a large space on the 2nd floor is a shared office with exposed beams. Different businesses, such as a design studio, a custom-made patisserie, a sweets cooking class and a CD shop, stand in a row in the 2 warehouses. A public space with grass was formed at the center of the premises and a place with lots of flexibility, which can be used as a party event space, was created.

We were able to find more historical buildings in this area and creators and the like renovated these buildings, creating various living spaces. Those people with an eye for detail gather in these places to "renew" and utilize. The place for interaction satisfies an intellectual curiosity and the number of these kinds of places has been increasing year by year.

　長野市の中心に鎮座する大寺院・善光寺の南側に位置し、かつて市内一の繁華街だった権堂町。そこに築100年ほどの2階建の屋敷と2つの蔵が立ち並ぶ、呉服問屋兼住居があった。建物が壊される際に「あまりに惜しい」と声があがり、ほどなく複数の借り手が現れ、入居者が協同で建物を運営するルールのもと、『権堂パブリックスペースOPEN』が開設されたのである。

　内部は入居者の手により多様にリノベーションされている。屋敷1階はカフェ、大空間の2階は梁むき出しのシェアオフィスに。2つの蔵にはデザイン事務所やオーダーメイドパティスリー、お菓子教室、CDショップなど異業種が軒を連ねる。敷地中央には芝生広場も設けられ、パーティ・イベント会場にもなる自由度の高い場が生み出されている。

　この地には他にも古い建物が数多く残り、そこにクリエイターらが改装を施し居を構えるケースが各所で見られた。精度の高いアンテナを持つ人々が多く集まり、そして彼らの手によって「更新」され活用される、知的好奇心を満たすサロンのような場が、年々その数を増し続けているのである。

アーツ千代田 3331
3331 Arts Chiyoda

学校の面影が利用者の記憶を刺激する
アートを媒介とした交流の場

An image of school to stimulate the memory of users
Interactive place with art mediation

Design
佐藤慎也＋メジロスタジオ
（黒川泰孝、馬場兼伸、古澤大輔）
Shinya Sato + Mejiro Studio

DATA
Site: Chiyoda-ku, Tokyo, Japan
Principal use: Art and Cultural Facilities
Total floor area: 2086.48㎡
Completion: 5/2010
Contractor: SAITO CONSTRUCTION COMPANY Ltd.

3331。この不思議なネーミングは、「江戸の一本締め[*1]」のリズムに由来するという。ここは閉校となった中学校をリノベーションし利用する参加交流型アートセンターで、行政の文化芸術振興プロジェクトにより誕生した。

隣接する公園と学校をウッドデッキでつなぎ、全面ガラス張りにした開放的な一階部分は、元は職員室だったというコミュニティスペースだ。ここは誰でも自由に利用でき、「お弁当持参で来る人も多い」とのこと。ガラス戸は全面開放可能で、夏にマルシェイベントが行われたこともあるという。

2・3階と地下1階は元の教室をそのまま生かし、入居者のアイディアによって手を加えられるギャラリーが入る。一部は間仕切りをしてシェアオフィスとして貸し出され、国内外アーティストが集中して作品制作できる滞在型スタジオも開設、プラットフォーム的役割も果たしている。

2階の一部は体育館になっており、近くの勤め人が仕事帰りに汗を流すことも多いという。そして屋上の元運動場には貸し菜園も設け、マーケットなどの大型イベントにも対応している。

公園との一体感を生む大きな開口部、そして誰もが親しみを覚える「学校」というハードの良い所をそのまま活用したことが、都市の中でひわだった求心力を発揮している空間構成上の一要因であろう。「何でもない廊下だけれど、なぜだか絵になる」。カメラマンはそう呟き、廊下とその曲がり角にある手洗い場に向け集中的にシャッターを切った。随所に残された学校建築の名残は、かつての子どもたちの遠い記憶を呼び起こさせる装置にもなっている。

1. 日本の伝統的風習として、ある出来事が滞りなく完了したことを祝し、当事者同士が掛け声とともにリズムを合わせて打つ手拍子のこと。手締め、手打ち。地域の祭りや式典、宴会の締めくくりなどに行われる。

3331 – this strange naming is derived from a rhythm of "Edo Ipponjime[*1]". 3331 Arts Chiyoda is an art center where people participate and interact with each other and it was established as a project for the promotion of culture and the arts. A closed-down junior high school was renovated and used practically for this art center.

The school building and adjacent park were connected with a wooden deck and the glazed open first floor, which used to be a staff room, is a community space. Anyone can use this space freely and a lot of people bring their lunch. Its glass doors can be opened fully and a market event has been held in summer.

The 2nd and 3rd floor and the 1st floor of basement still have the original classrooms and they have galleries where tenants can arrange with their artistic ideas. A section of these floors has partitions, which is rented out as a shared office space. In addition, a studio has opened where domestic and oversea artists can stay and focus on their works. The art center functions as a platform for artists.

A section of the 2nd floor is a gymnasium and a lot of people working in the nearby area come and do some exercise after work. In addition, renting vegetable garden is on the rooftop where a playground used to be and this space can also be used for big events, such as markets.

A factor of structuring this space structure to provide an outstanding centripetal force in the city was to use the good points of the material objects; a large opening that provides a sense of unity with the park and "school" with which everyone can feel familiarity. A cameraman muttered, "It is an ordinary corridor, yet somehow it makes a good picture", while he intensively snaps photos down the corridor towards a bathroom around the corner. Remains of school architecture, which can be found everywhere, serve to recall our old memories from when we were kids.

1. Hand clapping performed in a rhythm by the parties with shouts to celebrate a conclusion of an event without any delay is a Japanese traditional custom. It is one of the Tejime and Teuchi hand clapping styles. It is performed at the conclusion of a local festival or ceremony and at dinner parties.

SCAI THE BATHHOUSE

DATA
Site: Taito-ku, Tokyo, Japan
Principal use: Gallery
Total floor area: 244㎡
Completion: 1993
Structure: S structure

公衆浴場というユニークな場の再活用で
文化の街に加えられた多義的な刺激
Equivocal stimuli added to a cultural city by reutilizing a unique place, the public bath

Design
宮崎浩一
Mz design studio

東京都台東区谷中は、江戸時代から寺町として栄えてきたエリアで、戦災を免れ、昔ながらの街並みを残している。周辺には東京国立博物館や東京芸術大学などもある、文化的に恵まれた立地といえる。

　その地で200年にわたって営業を続けてきた銭湯「柏湯」は、1991年に惜しまれつつもその歴史に幕を閉じることとなった。建物の新たな使い道を模索するなか、この場で演劇上演会が行われたことを機に、文化的空間として活用することが提案される。1951年築の建物を改装し、「SCAI THE BATHHOUSE」が誕生したのである。

　入母屋造の瓦屋根の目を引く特徴的なファサードといい、下駄箱が並ぶ入り口といい、銭湯の名残は至る所に見られるが、ひとたび展示空間に入ると、作品を生かす白塗りの壁面が際だつ。高い天井から燦々と注ぐ自然光も、展示を生かす演出の一つとなっている。ここは最新鋭の国内アーティストを世界へ発信すると同時に、海外の知られざる作家を紹介する橋渡しを担う現代美術ギャラリーになったのだ。

　かつて、この国の代表的コミュニティ空間であった「銭湯」という場が、時を経て形を変え、人々をつなぐ役割を今なお果たし続けている。

Yanaka, Taito-ku in Tokyo is an area that flourished as a Teramachi (temple town) from the Edo period. The town was spared from war damage and still has a traditional townscape. It is also a culturally blessed location because the Tokyo National Museum, Tokyo University of the Arts and the like are in the surrounding area.

　The "Kashiwayu" public bath had been in business for 200 years in Yanaka but it unfortunately came to an end in 1991. A new use for the building was sought and, while a theatrical performance was held at the building, it was suggested that the building be turned into a cultural space. A building which was constructed in 1951 was renovated and "SCAI THE BATH HOUSE" came into being.

　The remains of the public bath, such as a characteristic facade drawing attention to the tiled hip-and-gable roof structure and an entrance lined with shoe cupboards, can be seen everywhere. Once we enter the exhibition space, white-painted wall surfaces allow art works to stand out. Natural light falling down brilliantly from a high ceiling is also one of aspects that fully takes advantage of exhibitions. This place highlights the latest domestic artists to the world and at the same time is a contemporary art gallery which introduces overseas artists little known in Japan.

　The "public bath", which represents community space in Japan, changes its form after many years but still plays the role of connecting people.

新宿マルイ 本館　屋上庭園「Q-COURT」
Shinjuku Marui Main Building, rooftop garden "Q-COURT"

都心の空と心地よい緑風を独占できる
屋上庭園という都市のオアシス

An oasis in the city center in the form of a rooftop garden
with the exclusive skyline of the city and a pleasant breeze through the greenery

DATA
Site: Shinjuku-ku, Tokyo, Japan
Principal use: Rooftop garden
Completion: 2009
Design: Emiko Futami (E.M.I.PROJECT)

かつてデパートの屋上の利用法といえば遊園地やビアガーデンなどが主流であったが、そのような「直接的な営利優先」の手法から、シャワー効果をねらった屋上緑化や、環境への配慮や社会的貢献にも応える手法をとる傾向が、時と共に一般化しつつある。ここ新宿マルイ本館屋上にも「都会の真ん中で、都会人に心からほっとしていただけるような場を」というコンセプトから、本格的な英国式庭園が設けられた。

　開放的な芝生広場を中心に、東西南北の各コーナーには個性的で居心地の良いソファーベンチや椅子が用意され、くつろぎの空間を演出する。庭園目当ての来訪客も多く、年配から子ども連れまで客層の幅も広く、新たな顧客層の発掘にもつながっているという。都会の空を独り占めしているような感覚を持てる穴場的空間として、家・職場以外の、人々の新たな居場所をつくるという試みがここに結実されている（49頁に記事）。

Rooftop areas in the past were mainly used for amusement parks and beer gardens, however, there is a general trend as time goes by to install areas such as rooftop greeneries, for a shower effect, and to be considerate of the environment and to make social contributions. Here, at the rooftop of Shinjuku Marui Main Building, a full-scale British garden is built with the concept of "a place in the city center for city people to relax from the bottom of their heart".

With a public space with grass in the center, in all four corners of the compass, individual and comfortable benches and chairs are installed and provide a relaxing environment. There are many people who come to visit the garden, and the types of visitors vary from the old to families with children, which are also leading to discover a new range of customers. An effort to create a new space other than home and work for people that provides a hidden space with a feeling owning the city sky is coming into fruition here (see p. 49 for details).

目黒天空庭園
Meguro Sky Garden

都市の異物になりがちな巨大建造物を
憩いの場へと変容させた天空庭園

A sky garden that transformed a massive building that could seem foreign in the city into a place of relaxation

DATA
Site: Meguro-ku, Tokyo, Japan
Principal use: Rooftop garden
Site area: 7,096㎡
Completion: 3/2013

　ここは首都高速道路大橋ジャンクションの屋上。地上高7〜35mにつくられた、ループ状の屋上庭園である。
　その巨大さ・排ガス・騒音などから「迷惑施設」とされがちな、高速道路ジャンクションという土木構築物。ここでは道路をコンクリートの壁で包み込み排気ガスや音を遮断する覆蓋構造が対策として採用された。その構造スケールからはローマのコロッセオが連想させられる。

　そして屋上には、四季折々の自然や和の文化を楽しめる回遊式の和風庭園が整備された。道路構造物に囲まれている内部空間には人工芝による多目的広場である「オーパス夢ひろば」が設けられ、合わせて活用されている。
　緩やかな勾配をつけられた坂を行く歩行路は周遊に心地よく、高さが変わるたびさまざまな眺めが展開される。東京の新たな視点場がここに生み出されている。一画には菜園も設けられ、天空庭

This is on the rooftop of Ohashi Junction of Metropolitan Expressway. This is a rooftop garden in the shape of a loop at 7 to 35 meters high above the ground level.

Junctions of expressways are civil engineering structures often seen as "nuisance" due to their massiveness, exhaust gas and noise. At this junction, a covering structure was adopted as a measure to cover the expressway with concrete walls in order to shut off exhaust gas and noise. This structural scale evokes an image of the Colosseum in Rome.

On the rooftop, a Japanese circuit style garden was established, where nature and Japanese culture in the four seasons can be enjoyed. Within the inner space surrounded by the structures of the expressway, "Opus Yume Hiroba", a multi-purpose public space with artificial grass is established and is well used.

The path on a gradual slope is suitable for walking around, and it also provides varied views from different heights. New viewpoints overlooking Tokyo are generated here. There is a vegetable garden in a corner, and volunteer workers of the garden

園で活動するボランティア団体が集い土を耕す。隣接する高層マンションともデッキで繋がれ、図書館などの公共施設も利用しやすく、複合的な都市空間が演出されている。

　アクセスはエレベーターや階段が主であるため、とうぜん都市との接続性という面からすれば地上の公園と同列には比較できないが、そのぶん素晴らしい眺望、開放感、そして何より他にはない「非日常的体験」を得られる場である。迷惑施設対策＝「臭い物には蓋を」という発想だけにとどまらず、より積極的な空間活用がなされたという点で、ここが数ある都市公園の中でもユニークかつ希少で、貴重な先例になっていることは疑いない。

gather plowing in the patch. The garden is connected to adjacent high-rise residential buildings with a decking, and community facilities such as a library are within easy access, which creates a multi-purpose urban space.

The access to the garden is mainly by elevators and stairways, therefore this garden cannot be compared to parks on the ground from the viewpoint of continuity with the city. However, this garden has an advantage to gain exceptional views, a sense of openness and an unparalleled "experience of the extraordinary". By going beyond the idea of "covering the problem" as a measure for a nuisance, the idea of actively utilizing the space is with no doubt making this park a very unique and rare, and a precious precedence for urban parks.

丸の内ブリックスクエア　三菱一号館広場
Marunouchi BRICK SQUARE

都心のポケットパークは
緑溢れる憩いの空間

*A pocket park in the city center is
a relaxing space full of greenery*

DATA
Site: Chiyoda-ku, Tokyo, Japan
Principal use: Public square
Site area: approx. 1,160㎡
Completion: 2009
Design: Mitsubishi Jisho Sekkei

　コンドル設計の明治時代の姿を忠実に復元した赤レンガの三菱一号館と、商業施設である丸の内パークビルディングが並ぶ丸の内の新空間「丸の内ブリックスクエア」。一号館広場はその一画に設けられた緑いっぱいの憩いの場である。メインゲートをくぐった先に広がる奥まった空間には、それ一本でシンボルツリーになるような多くの樹木が立ち並び、色濃い緑を演出。ベンチ以外にも腰を下ろせる場が各所に設けられ、利用する人が後を絶たない。

　建築物との関係性を考慮し、曲線を多用した落ち着いたデザインで「都心で心から憩える場を」というコンセプトに応えている（45頁に詳細記事）。

"Marunouchi Brick Square" is a new space in Marunouchi, with the red brick Mitsubishi Ichigokan that faithfully restored the original building designed by Conder in the Meiji Era, and is nearby the Marunouchi Park Building, a commercial facility. Ichigokan Square is a relaxation space full of greenery in a corner of Marunouchi Brick Square. After passing through the main gate, one enters a wide secluded space filled with a number of trees that are magnificent enough to serve as "symbol trees" providing a lush greenery.

There are spots for sitting down as well as benches, and visitors to the square are endless.

By taking the relationship with the architecture into consideration, the square's calm design makes great use of curved lines, and is aligned with the concept of "a space in the city center for full relaxation"(in-depth story in p. 45).

虎ノ門ヒルズ　オーバル広場
Toranomon Hills OVAL PLAZA

DATA
Site: Minato-ku, Tokyo, Japan
Principal use: Public square
Completion: 6/2014

車から人へ。
「立体道路制度」を活用した都心の芝生広場

From automobiles to people.
A public space with grass in the city center utilizing "Multi-Level Road System"

　2014年に開業した超高層複合タワー「虎ノ門ヒルズ」。道路の上下に建築物を建設でき、土地の有効利用を促す「立体道路制度」を活用して、建物内に地下トンネルを貫通させ、地上部は歩行者と緑に開放。訪れた人々がゆっくりと散歩や日光浴を楽しめる空間が提供されている。

　広大な芝生が広がるオープンスペースの一角にあるオーバル広場は、多種多様なイベント会場としても使われている。道路から一段高いことから開放感が生まれ、併設する屋外カフェも心地よく、足下のトンネルに車が走っているとは誰も思わないだろう。都市型広場の好事例である。

"Toranomon Hills" is a super high-rise multi-purpose tower opened in 2014. Making use of "Multi-Level Road System" allowed the effective use of land by constructing the building over and below the streets which provides an underground tunnel passing through the building and the ground space is made available for pedestrians and greenery. This place provides a space where visitors can enjoy strolling and sunbathing.

　Oval Plaza sits in a corner of the open space with an extensive grassed area and is also used as venue for various events. Being slightly higher than the street creates a sense of openness, and open-air cafes in the premises are so comfortable that nobody would think that automobiles are running in the tunnel below. This is a great example of urban public space.

都市の余白
A blank space in the city center

　超高層ビルのふもとにある半円形状の広場。周囲には彫刻が置かれ花壇が設けられ、訪れる人々の交流を促す場をと意図されたという。その見た目はカトリックの総本山、サン・ピエトロ大聖堂前の広場のようであり、それが象徴性を与えるのにふさわしいデザインと考えられたのだろう。

　ただ、そのスケールに比べ場に留まる者は少なく、閑散とした印象は拭えない。無機質な表情は各所に植えられた木々や花、彫刻で緩和されるが、やはりこの場の主役は背後にそびえ立つ高層建築であり、小さな人間は身体感覚を狂わされ、ただ圧倒される。

　ルールの多い囲いの中の遊び場では、子どもは次第に飽きる。その場のうまい活用のしかたを自分で発見したとき、子どもの目は最も輝く。一方、禁止事項だらけの公園など「安全の確保」が盾にされ、多様性を受け止められるキャパシティをもった「余白ある場」は都市から姿を消しつつある。管理下の安全性と自由度の高さをいかに両立させるか、都市が活力を戻す鍵はそこにあるのかもしれない。

A semicircular public space at the foot of skyscrapers. With sculptures and flower beds around the open space, it was intended to create a space to promote communication between visitors. Its appearance is similar to a square in front of the holiest site of the Catholic tradition, St. Peter's Basilica, and its design was thought to add appropriate symbolic character.

However, despite the scale of the open space, not many visitors stay there, and it is difficult to rid the impression of quietness. The coldness of the space is alleviated by the planted trees and flowers in various spots and sculptures, however, the main role of this space is the towering skyscrapers in the background, and hence the smallness of human beings loses their physical senses and only become overwhelmed.

Children gradually become bored in the gated playground with many restrictions. The eyes of the children shine most when they find themselves in a usable space. In contrast, "ensuring safety" is like a shield in a park full of don'ts, an "empty space" with a capacity to accept diversity is gradually disappearing from city centers. The key to regaining vitality in city centers may lie in finding the balance between safety under supervision and greater freedom.

興福寺　薪御能（薪能）
The Kohfukuji Temple, Takigi-Onoh (Takigi-Noh)

かがり火が照らす幽玄の広場
A public space with subtlety and profoundness under torchlight

DATA
Site: Nara-city, Japan

　主に野外に設置された舞台で繰り広げられ、照明として、また御神火として薪が焚かれたことから薪能と呼ばれるこの能は、奈良の興福寺南大門前の芝生を舞台とし、平安時代中頃に初めて催された神事がその発端と伝わる。明治期の中断を経て戦後、この御神事能は奈良県薪能保存会の主催のもと、「薪御能」として古儀に則り現在まで継続して執り行われている。そして、その薪御能をルーツとし、その形態を倣って行われているのが、全国各所の鎮守の森や城郭などを舞台に、長年にわたって賑わいを集め続ける薪能の催事なのである。

　徳川幕府により武家の式楽（公儀に用いる芸能）と規定され、明治維新と第二次大戦により存続危機を迎えたものの、その後大衆も楽しめる芸能として復興を遂げた能楽。身近な場を舞台とする薪能が、多くの人にその芸術的真価を知らしめる契機となった面は多分にあるといえよう。もしくは逆に、人々の渇望が薪能を引き寄せたのか。いずれにせよ、各所に出現するこの仮設性の高い広場で行われる催事は、その地における風物詩として人々を魅了し続けている。

This Noh was usually performed on an outdoor stage, and wood fire was used as sacred flame, which explains the origin of the name Takigi-noh (Torchlight Noh). This Noh is performed on the grass in front of the Southern Main Gate of the Kohfukuji Temple, and rituals first performed in the mid-Heian period are said to be the beginning of the Takigi-noh. After ceasing in the Meiji period, this Noh is now conducted by the Nara prefectural Takigi-noh preservation society as "Takigi-Onoh" in accordance with ancient rites. Takigi-noh events held nationally at sacred woods and castles have their roots in Takigi-Onoh and are modeled after the original style, and have been popular for a long time.

　Noh dramas were defined by the Tokugawa Shogunate as Samurai arts for formal occasions, and although it faced a crisis of continuance due to the Meiji Restoration and World war Ⅱ, it was revived later as a popular art. It is fair to say that Takigi-noh, which is based on ordinary scenes, was a key factor in showing its true artistic value to many people, or perhaps the opposite is true, that it was the people who yearned for Takigi-noh. Either way, the Takigi-noh events held at temporary public spaces in many locations continue to attract audiences as a seasonal attraction of the locale.

旧金毘羅大芝居（金丸座）
The Former Kompira Oshibai (Kanamaruza)

現存する国内最古の木造芝居小屋
歌舞伎上演で地域が一つになる

The oldest existing wooden theatre in Japan
The community is united by the performance of Kabuki

DATA
Site: Kotohira-cho, Kagawa, Japan
Principal use: Kabuki theater
Total floor area: 1,161㎡
Since: 1835

　ここは金刀比羅宮（119頁）の門前町に建てられ、今も現役で利用される、現存する日本最古の劇場型木造芝居小屋である。

　江戸中期、金刀比羅宮に詣でる金比羅信仰が全国的人気を呼び、門前町に参詣客の求めに応じて芝居小屋が常設された。明治期以後は一時映画館にもなったが、1970年に国の重要文化財の指定を受け、その後移築復原された。1985年からは、春に歌舞伎の定期公演が行われている。

　2004年の大改修により、竹格子の天井から花吹雪を散らす「ぶどう棚」と、役者が宙乗りするための「かけすじ」という仕掛けが復原された。舞台中央には人力で廻す「廻り舞台」を備え、江戸時代当時の芝居小屋の風情を今に伝える。舞台照明は今も自然光が中心で、舞台と客席の距離も近く、自然と役者と客の一体感が生み出される。公演は、地元民らのボランティアに支えられているという。今に残る貴重な建築遺産が街の活性化を促し、新たな人の輪を生み出している。

　The oldest existing wooden theater in Japan which was built in the shrine town of Kotohira-gu (page 119), is still used today.

　In the middle of Edo period, faith in the Kotohira Deity to visit Kotohira-gu became nationally popular, and drew visitors to Kotohira-gu so a theater was permanently built in the temple town to meet the demand of visitors. The theater once turned into a cinema after the Meiji period, however, it was registered as a nationally important cultural property in 1970, was later moved and restored where it remains until today. A regular Kabuki show has been performed there every spring since 1985.

　During the major renovation work in 2004, the devices, the "Budou Dana" (grapevine trellis), which scatter petals from the bamboo lattices on the ceiling, and the "Kakesuji" (rope and dolly system), which the actors use to fly in air, were restored. The "Mawari Butai" (revolving stage) which can be rotated by manpower was installed in the center of the stage, and as such, a taste of theater from the time of the Edo period is carried down to the present. The theater is still mainly lit by natural light, and the stage is close to the audience, which creates the united feeling among nature, actors and the audience. The performance is supported by volunteers from the local community. This existing architectural heritage is promoting revitalization of the town and is creating a new circle of people.

門前町（善光寺）
Cathedral town (Zenkoji)

単なる通過点ではない道。
「日本の広場」を象徴する道行き空間

A street that is not a simple passing point.
A street space that symbolizes the "Japanese Hiroba (public space)."

DATA
Site: Nagano-city, Japan

「遠くても一度は詣れ」。創建以来1400年余り、老若男女、またあらゆる宗派の人を平等に受け入れる無宗派寺院として庶民の信仰を集める信州・善光寺。全国から参詣客が集い、その表参道には参詣客を目当てに露店が軒を連ね、仮設だった露店が次第に常設・固定化していく。これが門前町のはじまりである。日本の多くの大都市が「城とその城下町」を基盤とするのに対し、ここは善光寺の門前町の発展が都市を形作った、国内でも希有な「門前町の都市」なのである。

参拝の到達地点である寺社の参道や門前町が栄える、これが西欧的広場を持ち得なかった日本が代わりに獲得した「広場」である。往来そのものが広場であり、そこが単なる通過地点でなく、通ることで心身を清め、最奥の聖地に近づく準備を整える。そんな儀式的な意味も備えた空間が形成されている。

"Although far, one must visit at least once." For over 1,400 years since its foundation, Zenkouji Temple has attracted faith from ordinary people as a non-sectarian temple and accepts everyone equally, young and old, and of any religion. In the old times, visitors milled about and stalls for visitors sprung up, opening side by side, and then temporary stalls became permanent and stabilized. This was the beginning of Monzenmachi (a temple town). Although many major cities in Japan are based on "a castle and its Jokamachi (a castle town)", this city is a rare "cathedral town", having emerged from the development of Zenkouji's temple town.

The approach to the shrines and temples, which are visitors' destinations, flourished as well as temple towns and shrine towns. These places became Japan's "Hiroba" instead of open spaces in the western notion that Japan was not able to achieve. Place for coming and going, or an approach itself is an open space, and not a simply a passing point. Going through the approach purifies one's mind and soul, and readies oneself to approach the most inward sacred location. A space with such ritualistic meaning formed here.

参道（金刀比羅宮）
The approach to Shrine (Kotohira-gu)

詣でる人々、その「ハレの日」を祝福する場
Visitors and a venue to celebrate the "grand day"

　最奥の聖地である本宮をめざし、つづら折れで先行きの見えない、700段を超える山道の階段を登り続ける。「こんぴらさん」の名で知られる香川県琴平町の金刀比羅宮への参詣は、江戸時代、伊勢神宮参りと並んで庶民の憧れであり一大行事であったという。今も毎年20万人以上の人出が集まり、その参道は賑わう。

　表参道の長い階段を登ることは、ケガレを払う禊ぎの行為そのものでもある。道中に本宮が見えないことは関係ない。「神聖なる方向」が示され、そこに身体を向かわせる行為自体に意義があるといえる。そして登山にも近い長い階段登りは、ある種のアミューズメント性も備えている。「詣でる」という行動様式は、ハレの行為であり大きな娯楽でもあるのだ。参道を華やげる仲店は、それを祝福する機能も果たしている。

DATA
Site: Kotohira-cho, Kagawa, Japan

People who aim to reach the most secluded main holy site, keep ascending over 700 steps of the mountain path with many switchbacks that are impossible to look through. Kotohira-gu shrine, known as "Konpira-san", in the town of Kotohira-cho in Kagawa, was in the time of the Edo period, a place ordinary people longed to visit as much as Ise Jingu shrine and visiting it was a once-in-a-life-time experience. Even now more than 200,000 people visit the shrine annually, and the approach to the shrine bustles with visitors.

　The act of climbing the long steps in the approach itself is also an act of cleansing oneself of impurities. It does not matter if the holy site does not come into sight throughout the approach. The fact that "the holy direction" is shown and the act of directing the body itself is significant. In addition, the long steps that resemble a mountain climb also provide a certain type of amusement. An act of "visiting" is an act of celebration as well as great amusement. The stalls livening up the approach also serve as a means of celebration.

鬼子母神　手創り市
Kishimojin, Marche

「境内の仮設市場」という
日本的な広場の現在進行形

*An evolving Japanese Hiroba (public space)
in the form of a "temporary market at a premise of shrine"*

DATA
Site: Toshima-ku, Tokyo, Japan

　安産・子育ての神として広く信仰を集める東京・雑司ヶ谷の鬼子母神堂。その境内では毎月一度、日曜日に「手創り市」が開催されている。その日は鬼子母神と近隣の神社境内に多くの露店が並び、客を呼ぶ。ハンドメイドの品を製作者自ら販売するこのようなマーケットは近年全国各所で開かれているが、この市は2006年頃から開催されており、東京における手創り市の草分け的存在である。

　「お客さんとの距離が近くて、品物のことだけでなくあれこれ話せるのがいい」と出店者は語る。物を売買する楽しみだけでなく、そんなコミュニケーショ

ンや賑わいそのものといった、「場を楽しむ」ために ここに集まる人も多いのだろう。売り物を地べたに並べる人、机や段を用意する人、ハンモックチェアで指圧マッサージを施す人など、出店者は与えられたスペースを生かして独自の販売空間をつくり、境内を賑やかにする。

　古来人が集まる場であり、そして人が集まる場所に設けられる「市」。特別な日に寺社境内に設置され賑わいを集め、そして終了後はすみやかに撤去される。扱う物は変われど、この国で何百年も昔から行われ続けている営みの原型がここに見てとれる。

Kishimojin Shrine in Zoshigaya Tokyo, is widely known as a goddess for easy childbirth and child rearing. A "handmade market" is held once a month on a Sunday on its premise. On the day of the market, the premises of Kishimojin and other neighboring shrines become crowded with stalls and visitors. Markets for craftsmen to sell their handmade items are held at many places nation-wide in recent years, and this market has been held since around 2006, and is a pioneer of handmade markets in Tokyo.

　One of the stallholders said, "I like it because of the closeness with the customers and for the opportunity to chat with them not only about my products but about anything". Many visitors must gather not only for the joy of shopping, but also for the opportunity to enjoy communication and the bustle and to "enjoy the scene". Stallholders create their shops and make good use of the limited space by sitting on the ground with their products, setting up tables and shelves, or by providing massages to people on a hammock chair, which add to the bustle of the area.

　Since the old times, shrines have been a place for people to gather, and this "market" is held at such a gathering place. A market is set up on special days within the premises of the shrine, attracts visitors, and dismantled immediately after the market is finished. Although different items are traded at the market, an original form of activity that has been conducted for the past several hundred years in this country is observed here.

花園神社　酉の市

Hanazono Shrine, "Tori-no-ichi"
(Festivals of Ohtori Shrine)

賑わいの磁力を発生し続ける
都心に出現した聖なる広場

A holy public space emerged in the city that continues to be a magnet for bustle

DATA
Site: Shinjuku-ku, Tokyo, Japan

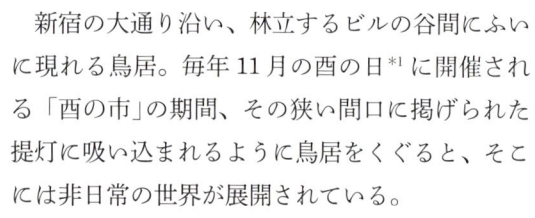

　新宿の大通り沿い、林立するビルの谷間にふいに現れる鳥居。毎年11月の酉の日[*1]に開催される「酉の市」の期間、その狭い間口に掲げられた提灯に吸い込まれるように鳥居をくぐると、そこには非日常の世界が展開されている。

　ここ新宿・花園神社の酉の市には毎年60万人以上の人が集い、飾り熊手の店や屋台が彩りを添える。また、独特の芸を売り物とする「見世物小屋」の興行が行われることでも知られている（2014年現在）。ここは長らく前衛的な演劇活動の拠点の場でもあるという。この場自体に、人間の潜在意識に訴える何らかの力が在るのかもしれない。

　祭りの夜、数多の提灯が境内を煌々と照らす。神の領域を借り行われる祝祭。聖俗が一体となって現出する場だ（39頁に記事）。

A Torii (shrine archway) unexpectedly appears on the main street in Shinjuku, between towering buildings. If you go under the Torii during the "Tori-no-Ichi" [*1] (Fair on the Day of the Rooster) held on the day of Tori every November, it is as if you are being invited by the lanterns that hang under the small archway, and that an extraordinary world awaits you.

More than 600,000 visitors flock to Tori-no-Ichi at Hanazono Shrine in Shinjuku every year, and shops to sell decorative Kumade ("bear-paw" lucky rakes) and other stalls add a nice touch as well. In addition, the festival is known to hold "Misemono Goya" (show booths) that shows unconventional entertainment (as of 2014). This venue also has long been a hub for avant-garde theatre activities. The venue itself may have some appealing power to one's subconscious.

On the night of the festival, a multitude of lanterns magically illuminate the grounds of the shrine. It's a celebration held at a borrowed sacred field. This is a venue where the sacred and the worldly emerge and come together (see p. 39 for details).

1.Days of each month are allocated for the twelve horary signs and the day of "Tori" here means the "rooster" days, in particular, the ones in November. Tori-no-Ichi are annual events held in many shrines, such as Ohtori-shrine in connection of "tori", dedicated to Yamato-takeru-no-mikoto.

1. 毎月の日付を十二支に振り当てて数え、その「酉」に当たる日のこと。11月の酉の日に開かれる酉の市は、日本武尊（やまとたけるのみこと）を祭神とする鷲・大鳥神社など「鳥」にちなんだ各神社で行われる年中行事。

巣鴨地蔵通り商店街(「とげぬき地蔵」)
Sugamo Jizo Street Mall ("Togenuki Jizo")

特異な存在感を発揮し続ける
高齢者のアミューズメントスポット
Amusement spot for elderly people with a unique presence

DATA
Site: Toshima-ku, Tokyo, Japan
Site area : approx. 770m

　4のつく縁日の日、JR巣鴨駅からとげぬき地蔵尊(高岩寺)までの道中は高齢者で埋め尽くされる。高岩寺を中心とし、駅から都電の停留場付近まで延びる「巣鴨地蔵通り商店街」。ここには露店も含め、甘味処・食事処・高齢者向け洋品店など顧客層に適う品々を扱う多くの店が並ぶ。「おばあちゃんの原宿」と称される所以だ。各所には高齢者に配慮した休憩所も多く設置されている。五街道の一つである中山道だったこの道は、歩行者が多いこともあり駅からの一方通行となっており、また時間により自動車の通行は制限されている。

　とげぬき地蔵には、境内の香炉で焚かれる線香の煙を浴びると病気が平癒するという縁起があり、また「洗い観音」と通称される、自身の治したい部分と同じ部分を洗うとご利益があるといわれる観音像も、多くの高齢者を呼び寄せる。参拝客にとっては疫病平癒の願掛けをすることはもちろん、ここで仲間たちと買い物や食事を楽しむことも大きな目的なのだ。

On festival days, which are days with '4' in them, i.e. 4th, 14th, 24th, the roads from JR Sugamo Station to Togenuki Jizo-son (Kogan-ji) are filled with elderly people. Sugamo Jizo Street Mall spreads out from the station to within close proximity of the Toei streetcar stop, with Kogan-ji in the center. Including street stalls, many shops handling sweets, meals and clothes for elderly people, line up here to meet the needs of their customers. Therefore, it is called "Grandmother's Harajyuku". A lot of rest stations, specifically for elderly people, are located all around the area. There are many walkers as this road was originally one of the Go-Kaido roads (the five major roads), so it is only a one way street from the station, and also motor traffic is limited at certain times of the day.

There is an omen that says Togenuki Jizo cures disease when people are covered with the smoke from an incense stick lit in the incense lamp around the temple grounds. In addition, the existence of the Kannon image called "Arai Kannon" brings many elderly people together because it is said that by washing a body part on the statue, the same body part will be healed. For visitors, the main purpose in not only to pray for recovery but also to enjoy shopping and have a meal with friends.

People are attracted to the area because it is a comfortable size as well as having good access for traffic to Kogan-ji. Unexpectedly, younger generations including grandchildren, are attracted to visit this place as well as elderly people. Having such a specific city space filled with lots of energy is unique and you will only be able to experience it here (see p. 38 for details).

高岩寺の広すぎず狭すぎないスケール感、往来からのアクセスの良さもその求心力を高める一因といえよう。高齢者だけでなくその孫世代など、若者も意外に多く訪れている。ほかでは味わえない、活気に満ちた特異な都市空間がここに提供されている（38頁に詳細記事）。

阿佐ヶ谷駅前広場
Asagaya station square

地域主体のジャズイベントが
街の日常を一変させる

*A community-run Jazz event
Transforms the daily life of the town*

DATA
Site: Suginami-ku, Tokyo, Japan
Principal use: Station square

JR中央線の阿佐ヶ谷駅前には、近隣のビルに負けず高く聳え立つ杉の木が点在する。都内の駅前広場にある木としては異例の高さで、街の玄関口のシンボル的存在となっている。そしてここは、毎年行われるジャズイベントの会場でもある。

　駅前広場から街の中心にかけて、夏にはトンネル状に葉が生い茂るけやき並木の通りが走る。駅前広場と並木の通りを軸に、神社・教会、公園、企業のロビー、喫茶店、レストランなど日常の生活空間が会場に利用され、街はジャズで埋め尽くされる。イベントの企画・運営はいずれも地域住民の手に担われている。「発想ひとつで街が一変する」という確かな実績が、地域にさらなる活力を与えている。

Cedar trees towering as tall as neighboring buildings are scattered in front of Asagaya Station of the JR Chuo Line. These trees constitute a grove in the station square, which is exceptionally tall for the urban context of Tokyo, and it is symbolically regarded as the gateway to the town. This site is also the venue for an annual Jazz event.

　A street lined with a row of zelkova trees runs from the station square to the town center, and their leaves create a green tunnel in the summer. Centering around the station square and the street, living spaces like shrines, churches, parks, company lobbies, cafes and restaurants are used as venues for the event, and the town becomes flooded with Jazz. Both the planning and implementation of this event are undertaken by the local community. The real achievement "to transform the town with a single idea" provides further vitality to the community.

ポンテ広場
PONTE SQUARE

賑わいの萌芽を
大らかな目で育むダンスの聖地

Sacred ground for dances
which nurtures the start of a crowd with tolerant eyes

DATA
Site: Osaka, Japan
Principal use: Public square
Site area: approx. 16,000㎡
Completion: 1996
Design, Management: Nikken Sekkei LTD.
Contractor: Kajima Corporation, DAITETSU KOGYO CO., LTD. (Joint Venture)

関西空港や全国主要都市を結ぶバスが発着し、JR難波駅と直結する「大阪ミナミ」の玄関口、OCAT（大阪シティエアターミナル）。その一画に「ポンテ広場」と呼ばれる地下一階の吹き抜けの広場がある。その底地は大阪市が所有、第3セクターが管理するなど、公民交じった運営形態が取られている。

　実はここ、この地域のストリートダンスの聖地なのである。ガラスなどを鏡代わりにするダンサーが集まり始めたのを知った管理側が「賑わいのきっかけになれば」と鏡を常設。ダンス練習の使用に際しては申請も不要とし、若者たちがさらに集まりやすい環境が整えられた。ガラス張りのビルなどが自然発生的にダンス練習の場にされる光景は都会でよく見かけられるが、ここはそれが公に許された希有な場なのである。ダンス大会やその他のイベントも定期的に開かれ、多くの人手を集める。それらイベントも管理者への申請で開催が可能と、柔軟な形で運営されている（一部制限あり）。

　近隣からの苦情もなく、むしろ貴重な居場所を守るべく、若者たちは率先してトラブル防止に努めているという。自由な環境から自治が生まれ、その自治により自由が保たれる。理想的な都市の公共空間が形成された場である。

OCAT (Osaka City Air Terminal) serves as a gateway to "Minami, Osaka," where buses connect with Kansai International Airport and with major cities in Japan, and which connects directly with JR Nanba Station. In one area of its basement, there is a public space with an open ceiling, called "PONTE SQUARE." It is run through the means of a joint management structure mixed with the public and private; Osaka-city owns the underlying land and the third sector manages it.

　The truth is this is a sacred ground for street dances in this district. The management side that learned of dancers gathering to practice using the glasses as mirrors, permanently installed mirrors with an intention to "boost the turnouts of crowds." They also permit usage without any application for dance practices to further enhance the environment where the youth can easily gather together.

　The scenes where glass-walled buildings are spontaneously made into dance practice venues can be often seen in urban areas. This is one of the rare venues where such scenes are welcomed by the public. Regular dance competitions and other events take place here to attract large crowds. Management is operated in a flexible manner where these events can be held with an application to the administrator. (Partial limitations apply.)

　No complaints have been raised by the neighborhood so far. The youth are rather actively making an effort to prevent troubles to protect their precious arena. The free environment creates autonomy; the autonomy maintains the freedom. This is a venue where an ideal urban public space has been established.

歩行者天国（銀座）
Pedestrian mall (Ginza)

車が消え、生まれる非日常
「街に参加する」感覚がもたらされる瞬間

Automobiles disappear and an extraordinary life appear
A moment of "being part of the town" is felt

DATA
Site: Chuo-ku, Tokyo, Japan
Site area: approx. 1,100m

の精神の現れともいえる。

　車を止める、それだけのことで日常が非日常に変わる。道行く人の表情や装いが晴れ晴れしく感じられるのは、車を気にせず歩ける快適さはもとより、自分もこの「ハレの場」を演出する一員だということに自覚的な人が多いからかもしれない。

In 1970 when Japan was in the midst of a high-growth period, a "pedestrian mall" was first introduced in Ginza, Tokyo. Currently not only automobiles but also bicycles are banned from passing for a few hours, mainly in the afternoons on weekends, and it serves as a street exclusively for pedestrians to safely stroll and shop.

　What makes the pedestrian mall in Ginza so characteristic is the fact that trash cans, parasols and chairs are set in the street. Having a space for relaxing under the scorching sun shows the welcoming spirit of the local community to provide as much a comfortable venue as possible to visitors.

　By a simple act of stopping the traffic, the daily life is transformed into an extraordinary life. Faces and looks of pedestrians appear in high spirits not only because of the comfort of being able to walk without having to worry about the traffic, but also because many of them are conscious that they themselves are one of the factors that create the "grand occasion".

　日本が高度成長期の只中にあった1970年、東京・銀座で初めて実施された「歩行者天国」。現在は主に休日の午後の数時間、車だけでなく自転車も走行禁止とされ、歩行者専用道路として、安心して散策やショッピングができる場が設けられている。

　銀座の歩行者天国が他のそれと比べて特徴的なのは、道路上にゴミ箱、パラソルとイスが設置されている点だ。炎天下でも休憩できるスペースが準備されているのは、訪れる人々に少しでも心地よい場を提供しようという、地域住民のもてなし

お台場海浜公園
Odaiba Marine Park

都市に造成された貴重な水辺
A precious waterside developed in city centers

DATA
Site: Minato-ku, Tokyo, Japan
Since: 1975

This site used to be a timber yard, and is now a man-made beach on a peaceful bay surrounding the remains of Daiba (a marine battery for coastal defense). With fusion of the sea, greenery, and the futuristic scenery of Rainbow Bridge, Fuji Television and urban resort hotels, this park is one of the most popular spots in Tokyo for sightseeing and dating.

The beach is an inverted L shaped shoreline. Although it is a man-made beach made of artificial sand filling, it is one the few sand beaches in Tokyo Bay and its existential value increases due to its convenient location in the central area of Tokyo Waterfront City. This precious waterside in the city center is enjoyed throughout the four seasons by a great number of people, including families playing in the water and couples watching the sunset.

　ここはかつて貯木場であり、海防のための海上砲台＝台場があった湾岸地。その台場跡を囲む静かな入り江の人工海浜である。海や緑にレインボーブリッジ、フジテレビや都市型リゾートホテルなどの未来的景観が融合し、都内でも観光地として、デートスポットとして高い人気を誇る。

　汀線が逆L字型の砂浜は、砂を運び入れて造成した人工海浜ではあるものの、臨海副都心エリアの中心部にあることで、東京湾内の数少ない砂浜としてその存在価値を増す。水遊びに興じる家族から夕暮れを見つめる恋人たちまで、貴重な都心の水辺は四季を通じて多くの人々に愛されている。

カナルカフェ
CANAL CAFE

100年をかけ熟成された「お濠の広場」
"A public space on the moat" that matured over a hundred years

　東京には魅力ある水辺は点在するが、見落としがちなのが外濠である。江戸城の遺構で、今は渡り鳥が降り立ち、亀や鯉が悠々と泳ぐ都心のオアシス。その水面にせり出すように建つ一軒のレストランバーがある。大正時代、東京初のボート場として設立された「東京水上倶楽部」が運営する「カナルカフェ」である。

　江戸情緒を残す貴重な憩いの場として長く利用されてきた東京水上倶楽部。時を経てボート利用者が減るなかで、「この歴史ある水辺を再び憩いの場に」と考えた現オーナーが創業したレストランは、席と水面がほど近く、他に類を見ない「水辺との一体感」が生み出されている。「人間にとって切っても切り離せない『水辺』を都市にいながら体感することができる希有な場を、広く皆様に開放したかった。赤ちゃんからお年寄り、そして世界中から集まる人々が心地よく過ごせる、ここは今、東京一幅広い層の人々が集う店かもしれないし、そうでありたいと願っています」とオーナーは語る。

DATA
Site: Shinjuku-ku, Tokyo, Japan
Since : 1918 (TOKYO SUIJYO CLUB)

設立から100年余、受け継いだ場を守るべく続けてきた水質浄化運動が奏功し、数年前からは蛍も舞うほどという。「都市の水辺を見直し活用する」という目的を一とする「エコ地域デザイン研究所」（陣内秀信所長）の発案で行われる水上コンサートの場も提供し、夏の夜、ボート上から、そして水上デッキから演奏を楽しむ多くの人を集めている。

100年の月日をかけて築かれた憩いの水辺は、次の100年を見据え、都市を潤し続ける。

Although Tokyo has a lot of attractive waterside areas, the outer moat is always a good surprise. This outer moat in the historical site surrounding the Edo Castle provides an urban oasis where migratory birds gather, and tortoises and carps swan. A restaurant stands reaching out to the water's surface. This is CANAL CAFE, run by the "Tokyo Suijyo Club," the first Tokyo boat house established in the Taisho period.

The Tokyo Suijyo Club has long been used as a precious relaxation spot with sentiments from the Edo Period. Over time, as the number of people who came to row boats declined, the restaurant was opened by the current owner who wanted to "make this historical waterside into a relaxation spot again". The seats of the restaurant are close to the water, which creates an exceptional sense of "unity with the waterside". The owner said, "I wanted to open this rare space to the public where the public can experience the 'waterside' in the city which cannot be separated from the life of humans. This restaurant may be, and I hope it is, a place for the widest diversity of people in Tokyo, from the young to old, and from all over the world, to relax."

It has been over one hundred years since the Club opened, and the water purification operation to maintain the heritage has been successful enough to have fireflies come back in the past few years.

Upon a suggestion of the "Laboratory of Regional Design with Ecology" (Director, Hidenobu Jinnai) with a purpose to reevaluate and revitalize waterside areas of the city, concerts are performed on the water by the restaurant, and crowds of people enjoy performances while sitting on a boat or on a deck on summer evenings.

The waterside for relaxation that came to establishment after a hundred years will continue to run in the city for another hundred years.

お鷹の道
Otaka-no-michi

古代から伝わる良質な水源が
地域の絆を育み続ける

*Quality water sources from the ancient times
nurture bond of the community*

DATA
Site: Kokubunji-city, Tokyo, Japan
Site area : approx. 350m

　東京都西部、多摩川が武蔵野台地を浸蝕してできあがった国分寺崖線といわれる段丘。その湧水を水源とする小川沿いに、かつてこの一帯が尾張徳川家の鷹狩り場であったことから「お鷹の道」と名付けられた遊歩道がある。

　縄文時代から使用されていたと伝わる豊富で良質な水場は、今も近隣住民が水汲みに訪れ、野菜を売る露店もある。休憩スペースでは顔なじみが弁当をつつきつつ談笑する。

　道脇に湧水園・植物園などが連なり、近隣には史跡地域の総合案内所として「史跡の駅　おたカフェ」が運営され、地域交流の拠点となっている。同所では水を切り口とした講座「水の学校」を開講するなど、貴重な資源を守り次世代に引き継ぐための活動が積極的に行われている。

Kokubunji Gaisen is a terrace created by erosion on the Musashino Plateau by the Tama River in the western Tokyo. Along the creek, flowing from a natural spring, is a path named "Otaka-no-michi" (path for hawks) as the whole area was once a site used for hawking by the Owari Tokugawa family.

It is said that this site produces quality water used since the Jomon Era, and even now people in the neighborhood come for its water, and there are some vegetable stalls there as well. Familiar faces enjoy lunch and chat at this relaxation space.

A spring park and a botanical park line the path, and the nearby "Shiseki-no-eki Ota Cafe" operates as an information center of historical sites, serving as communication hub in the community. This cafe also promotes activities such as "Mizuno Gakko" (water school) in order to preserve the precious resources so they can be enjoyed by future generations.

隅田川テラス
Sumidagawa Terrace

人と水辺を切り離さない堤防が
川沿いに穏やかな「開放区」を生む

*An embankment that does not separate people from the waterside
creates a serene "open space" on the riverbank*

DATA
Site: Chuo-ku, Tokyo, Japan

　オフィスや住居、ホテル、レストランなど多彩な機能を持つ複合施設「聖路加ガーデン」。その敷地内にある隅田川沿いの親水公園からは、河畔の素晴らしい光景を楽しめる。川に向かって設けられた半円形の階段はベンチとしても有効に活用されている。

　この一帯は、川の両岸に沿って整備された親水テラスと呼ばれる遊歩道である。かつて汚染もひどかった隅田川沿いには「カミソリ堤防」と呼ばれる塀のような堤防が築かれ、人々は川から遠ざけられていた。この状況を見直すべく、1980年代から、傾斜を付け川沿いのビルや道路と一体に開発し水際に近寄れるようにした「スーパー堤防」が整備されたのである。

　都市の発展のためいち早く犠牲にされた東京の水辺。そんな負の歴史を記憶に留める者は、もはや少なくなっているだろう。

"ST Luke's Garden" is a multi-purpose complex with varying functionalities such as offices, residences, hotels, and restaurants. The fantastic scenery of the waterfront can be enjoyed from the waterfront park along the Sumida River within the garden. The semicircular staircases facing the river are also used effectively as a bench.

This whole area is a path called waterfront terrace which was developed along both sides of the river. In the past, fence-like embankments called "razor embankments" were built along the then highly polluted Sumida River, and people were made to stay away. Since the 1980's, in order to improve the situation, slanted "super embankments" were established which enabled buildings and streets along the river to be developed together as a whole and they provide an approach to the waterfront.

Waterfronts in Tokyo were the first to be sacrificed when the city was developed. Only a few may recall such a negative history.

ネオ屋台村® 有楽町東京国際フォーラム村
NEO STALL VILLAGE®

賑わいが賑わいを呼ぶ
都心のランチタイムを彩る屋台空間
*With bustle calling for more bustle,
a stall space that enriches urban lunch time*

DATA
Site: Minato-ku, Tokyo, Japan
Site area: approx. 2,090㎡
Since: 2003

　平日のランチタイム、東京駅にほど近い有楽町の東京国際フォーラムの地上広場は、カラフルな移動販売車で埋め尽くされる。まるで東南アジアの屋台市場のような異空間がオフィス街の中心で展開される。
　「ネオ屋台村」と名付けられたこの販売活動は、首都圏の20カ所を超える地で営業を展開。移動販売事業者が運営事業者に出店登録し、自身

の移動販売車の中に厨房を作り、その場で調理してできたてを販売。リーズナブルな価格と質で多くの顧客を集めることに成功している。

　お洒落な販売車が複数台集結することで生まれる彩りや、楽しく賑やかな雰囲気が、人を引き付ける最大の演出効果であろう。たとえ好立地の広いスペースがあっても、きっかけがなければ賑わいは生まれない。良いサービスを提供すれば勝手に人は集まるはずだ、そんな単純で傲慢な考えを超えた運営側の視点があってこその賑わいだろう。賑わいそのものが人を呼び、人が集まってまた賑わいとなる、その好循環は続く。

At lunchtime on weekdays, ground public space at Tokyo International Forum in Yurakucho near Tokyo Station is filled with colorful mobile vending vehicles. An extraordinary space like a stall market in South East Asia is held in the middle of the business district.

　This market activity known as, "NEO STALL VILLAGE" is held at over 20 locations in the Tokyo metropolitan area. Mobile vendors register their stalls with organizers, and they sell freshly cooked foods in their mobile vehicles with fully equipped kitchens. By providing reasonable prices and quality, foods, they are successfully attracting many customers.

　Colorfulness created by several stylish vending vehicles lined up and cheerfulness in the bustling atmosphere may be the village's best dramatic impact in attracting customers. Liveliness does not bloom without a trigger, even though this is a wide space in a great location. The organizers' viewpoint goes beyond a simple and arrogant idea that the "customers will naturally come when a good service is provided", and it must bring in the bustle to the village. Bustle calls for customers, and customers create more bustles. This virtuous cycle continues.

太陽のマルシェ
Taiyo-no-marche

コミュニケーションのきっかけを人々に与える
都心の現代型市場

*A modern market in the city center
that provides people with triggers for communication*

DATA
Site: Chuo-ku, Tokyo, Japan
Since : 2013

　古くから隅田川を中心にベイエリアとして発展してきた勝どき周辺は、近年タワーマンションが続々と建設され、築地市場や2020年東京オリンピックの会場にも近く、商業やビジネスの中心地である銀座や丸の内へのアクセスも良い。今後さらなる発展に期待が高まるこの湾岸エリアをより活性化させるため、「食べる・買う・学ぶ・体験する」をコンセプトに、デベロッパーと地域町内会が主催し、日本最大規模の定期開催・都市型マルシェが企画された。旬の野菜や果物の販売に加え、親子で楽しめるワークショップやイベントなどを体験できる「太陽のマルシェ」である。

　色とりどりの移動販売車が集まる飲食エリアと、新鮮な野菜やワイン、加工品がならぶ物販エリア。試食を勧められ、その味に顔をほころばせる来場者。すぐ目の前に生産者がいるので、買い物ついでに最適の調理法を聞くなど、積極的なコ

Developed since the old times as a bay area around Sumida River, where in recent years tower apartment buildings have popped up one after another, this area is in close proximity to the Tsukiji Market and to the site of 2020 Summer Olympics in Tokyo, and has easy access to business centers such as Ginza and Marunouchi. With the purpose of further vitalizing the bay area, which is expected to develop even more, and with the concept of "Eat, Buy, Lean, Experience", this regular urban marche was hosted by the developers and the local neighborhood. In addition to sales of seasonal vegetables and fruit, people can experience workshops and events for families at the "Taiyou-no-marche".

Colorful mobile vending vehicles liven up the eat-in area, and fresh vegetables, wine and processed goods are lined up in the market. Visitors are offered sample tasting and smile at the pleasant flavors. Another a great attraction of the marche is that the producer right in front which triggers active discussion such as the best recipes for the product once purchased. Each market hosts approximately 100 stalls, each of which attracts around 10,000 people per day, and is a good example of a "modern market in the city center".

ミュニケーションを生み出すきっかけを与えられているのも大きな魅力だ。毎回約100店舗が出店し、一日に約1万人もの人を呼び寄せる、「都心の現代型市場」の好事例である。

橋詰広場（日本橋）
Bridgehead plaza (Nihombashi)

交通の要所に自然発生した
かつての「日本的広場」の代表

An example of former "Japanese Hiroba" (public space) that naturally appeared at the hinge of traffic

DATA
Site: Chuo-ku, Tokyo, Japan
Site area: approx. 50m
Since: 1603

　江戸幕府が開かれた1603年に架けられて以来、東海道をはじめ五街道の起点と定められ、主要水路である日本橋川との交差点としても重要な位置を占めた日本橋。そこにはかつて官許の芝居小屋があり魚河岸があり、それをとりまいて商人や芸術家などの人、またモノ・情報すべてが集まる文化的要所であった。

　橋梁技術が未発達だった当時、なるべく橋を短くするため岸から川に向けて張り出すように土台を築いたり、橋幅を道幅より狭くしたりする工法がしばしば取られ、その結果橋のたもとがふくらんで、必然的に橋詰には広い空間が現れた。これが「橋詰広場」である。多くの人・モノが往来する交通の要所である橋に付随するその空間が、いわゆる「西洋的広場」と同じような役割を果たすパブリックスペースとなったのも自然の流れといえよう。

　幕府にとってそこは、統治政策を周知させるための格好の場である。一方には法度や掟、罪人の手配書などを掲げる「高札場」が設けられ、幕府の意向を伝えるPRセンターとなった。そしてもう一方に設けられたのは、罪人の首を晒す「晒し

場」である。封建秩序を乱す者を処罰し、幕府権力を庶民に示す社会統治のシンボルとしてこの場が利用された。

　その後幾度かの変遷を経て、現在架かる日本橋は1911年に完成、国の重要文化財に指定されている。高札場・晒し場のあった広場には記念碑や花壇、水上クルーズの乗降場が設けられ、訪れる観光客も多い。東京を象徴する場として人々を引き寄せる一方、「憩いを得られるか」という面からいえば、水辺に近づける親水テラスのような施設がない点、高速道路が上空を覆い美観を損ねている点などの課題も残される。とはいえ、ここが象徴的な場であるゆえ、地域住民を中心にその改善をめざす動きは根強い。現代の多くの人々に望まれる「広場」として、新たな都市空間がこの地にお目見えする日も、そう遠い未来の話ではないだろう。

Since its erection in 1603 at the start of the Edo period, Nihombashi Bridge was the origin of the Five Routes including Tokaido, and played an important function as a crossing over the main waterway, the Nihombashi River. There were once official theaters and fish markets here and this site was a cultural center that attracted people like merchants and artists who came to the theaters and the markets for goods and information, among other things.

In those times, the bridge building techniques were underdeveloped. Construction methods, such as building a base stretching from the riverbank to the river and making the bridge width narrower than the road width, were often adopted to shorten the bridge as much as possible. As a result, the foot of the bridge was widened and a wide space was set up there.

　This space at the foot of the bridge formed the "bridgehead plaza". It is natural that a space attached to a bridge over a traffic hinge with heavy traffic turned into a public space and served similarly as a so-called "Western plaza".

This became an ideal place for the Shogunate to communicate government policies. One side, known as "Kousatsuba", was used to display acts, rules and search instructions of criminals, and was set up and served as a public relations center to make known the Shogunate's intentions. The other side, known as "Sarashiba", was used to display heads of criminals. This was used as a symbol of governance to show the Shogunate's authority to ordinary people by executing those who disturbed feudal peace.

After several changes, the current Nihombashi Bridge was completed in 1911, and is registered as an important national cultural property. Memorial statues, flowerbeds and stops for water cruises have been installed in the public space where Kosatsuba and Sarashiba once were, and many tourist visit here. While it continues to attract people as a symbolic place in Tokyo, it does not succeed in "providing relaxation" to people, due to some issues such as the lack of waterside terrace for an easy access to water or the interrupted scenery caused by overpassing highways.

However, there is a strong movement for improvement, mainly by the local community, as this site is such a symbolic place. It may not be so long before a new urban space will appear here as a "Hiroba" wanted by many people in the current times.

写真
Photo Credits

	鈴木知之　*Tomoyuki Suzuki*
pp. 18-21	① 藤塚光政（pp. 18-19、p. 20中）　*Mitsumasa Fujitsuka*
	② Erieta Attali (p. 20上・下、p. 21)
pp. 30-33	Nicolas Waltefaugle
pp. 45（図17）	画像提供：森ビル株式会社　*courtesy: Mori Building Co., Ltd.*
p. 49	画像提供：三越伊勢丹ホールディングス　*courtesy: Isetan Mitsukoshi Holdings Ltd.*
pp. 54-55	① 渡邉 修（p. 54上、p55下）　*Osamu Watanabe*
	② 中道淳／ナカサアンドパートナーズ（p. 54下、p. 55上）　*Atsushi Nakamichi / Nacása & Partners*
	画像提供：金沢21世紀美術館（①、②とも）　*courtesy: 21st Century Museum of Contemporary Art, Kanazawa*
pp. 56-57	西沢立衛建築設計事務所　*Office of Ryue Nishizawa*
pp. 62-63	画像提供：NAP建築設計事務所　*courtesy: Hiroshi Nakamura & NAP Co., Ltd*
p. 68	西川公朗　*Masao Nishikawa*
pp. 74-77	小野寺康　*Yasushi Onodera*
pp. 82-83	表 恒匡　*Nobutada Omote*
p. 84（上）	画像提供：森ビル株式会社　*courtesy: Mori Building Co., Ltd.*
pp. 86-89	藤村泰一　*Yasukazu Fujimura*
pp. 100-101	画像提供：権堂パブリックスペースOPEN　*courtesy: Gondo Public space OPEN*
p. 114	画像提供：森ビル株式会社　*courtesy: Mori Building Co., Ltd.*
p. 116	桑原英文　*Eibun Kuwabara*
p. 117	画像提供：琴平町教育委員会　*courtesy: Kotohira-cho, Board of Education*
pp. 118-119	編集部　*Tankosha Editorial Department*
pp. 128-129	画像提供：湊町開発センター　*courtesy: MDC*
pp. 152, 154	小野寺康　*Yasushi Onodera*

本書掲載の各事例の現況は、取材時（2014年12月現在）の状況と異なっている場合がある。
Each instance that appears in this publication is explained based on conditions at the time of coverage (as of December 2014), and may differ from present conditions.

掲載事例の所在地詳細はp164〜165を参照のこと。なお、表記は2014年12月現在の地名にしたがった。
Refer to pp. 164-165 for the address details of published instances. The addresses shown are consistent with the location names as of December 2014.

営業時間など各事例の詳細情報については、p164〜165に記載したURLにより確認されたい。
Refer to the URLs that appear on pp. 164-165 for the detailed information of each instance, such as opening hours.

広場を持たない都市

　広場、という都市文化を日本は歴史的に持ってこなかった。

　この場合の広場とは、plaza、piazza という類の欧州的な意味での広場、つまりカミロ・ジッテが『広場の造形』(13頁参照)で取り上げ、ポール・ズッカー(1888～1971　建築家・都市計画家・美術史家)によって『都市と広場』という大著が編まれた対象としての広場のことである。

　日本の伝統的な都市で、同様の活動が行われていた場としてそれに相当するのは、大路(広幅員の街路)や寺社の境内、あるいは名所と呼ばれる景勝地や河原、橋詰などであった。

　オギュスタン・ベルク(1942～　地理学者。1984～88の間、東京日仏会館学長を務める)も、「伝統的な日本の都市における広場の欠如」とともに「日本の都市の場合、西欧では広場で繰り広げられる活動が、一般に街路を舞台に行われる」ことを指摘した(『空間の日本文化』筑摩書房)。

　祝祭行事などが、街路のほか、寺社の境内や河原などで行われるという慣習は今日に続いている。新年に行われる火祭りとして「左義長」という祭事は全国にいくつか見受けられるが、福井県勝山のそれは、毎年1月に町中に(つまり街路に)「櫓」が出て踊りと音楽が供され、最後は「ドンド焼き」といって河原で飾り物を集めて焼き、その炎で餅を焼いて食することで、その年の五穀豊穣と健康を祈願する。日も暮れた河原で数々の松明が焚き上げられる中、長い竿の先に餅を付けて炎に差し込まれ、暗闇の中に人々の笑顔が照ら

Cities without Public Spaces

Historically, Japan did not have an urban culture of public space.

"Public space" is considered here in the European sense, such as plazas and piazzas. In other words, it is a public space which Camillo Sitte covered in his book, "City Planning According to Artist Principles" (See p. 13) and which is a subject in the voluminous work, "Town and Square" by Paul Zucker (1888-1971, architect, town planner, art historian).

In traditional Japanese cities, main streets (wide lane streets), premises in temples and shrines, famous scenic places, river beaches and bridge guards served as the corresponding places where activities were held as would be held in European public spaces.

Augustin Berque (1942- geographer 1984-1988 President of La Maison Franco-Japonaise) also noted the "lack of public spaces in traditional Japanese cities" and that "activities taking place in Western European plazas generally take place on streets in Japanese cities" ("Living Space in Japan", ChikumaShobo).

新たなパブリックスペースの復権
小野寺 康 (小野寺康都市設計事務所)

Restoration of New Public Spaces
Yasushi Onodera　(Office of Yasushi ONODERA – Civil Engineering & Landscape Architecture Design Office -)

し出されるというのは幻想的な風景である。

しかし、寺社の祭事がその境内で行われ、町の祝祭行事が河原で開かれたとしても、それは一時的なものであり、いわゆる常設された広場での活動と同じものではない。最近復刻された『日本の広場』（都市デザイン研究体・彰国社・37頁参照）では、このような場の出現を「広場化」と呼んでいる。日本には広場はなかった。広場化することでそれは存在してきた、というのがその論旨だ。

ギリシャ、ローマを起源とする西欧都市構造は、中心部にアゴラ、フォルムといった広場を建設し、そこを商取引などの市場や、政治や裁判の集会場、あるいは宗教的祝祭の舞台として稼働させた。まさに都市は、広場を造形することで都市たり得たのである。

中世期には、城塞に囲まれた自治都市が発展し、教会や修道院、あるいは市庁舎といった、精神的中核や政治経済の中心に寄り添う形で、広場は構築されてきた。ルネサンス期やバロック期以降は、次第にそれらを基点に都市全体を構成するロジックが成長し始める。

一方で伝統的な日本の都市の骨格は、このダイアグラムには乗らない。

槇文彦が著書『見えがくれする都市』（鹿島出版会）の中で、西欧の都市構造を「中心―区画」になぞらえ、日本のそれを「奥―包摂」と表現したことは示唆に富むが、西洋都市が中核施設に広場を付随させて「まちの中心」を形成し、それに収斂する形で街区が展開されているとするなら、伝統的な日本の都市構造は全く違うものだ。

城下町であれば城を基点として、部分的には格子構成（グリッドパターン）による町割りが散見される。しかし、城

Traditional events, such as festivals, were performed not only on streets but also at other places including premises in temples and shrines and river beaches and that is still the case today. The New Year fire festival "Sagicho" is held at several places in Japan. The fire festival is held in Katsuyama, Fukui, in January every year and "Yagura" (high wooden stages) are placed everywhere in the town on the streets for dance and music performances. At the end of the festival, ornaments are collected and set alight at the river beach ("Dondoyaki"). Then people pray for a bumper harvest for the year and health by eating rice cakes grilled on the fire. Rice cakes on long sticks are put into the fire and people's smiles are lit up in the dark while lots of flaming torches are lit at the river beach at dusk, creating a dream-like atmosphere.

However, even when the rituals of temples and shrines are held in their premises and the festival events of the town are held at the river beach, they are temporary and different from activities held in so-called permanent public spaces. The existence of such spaces, called "Hiroba-ka", in Japan is addressed in the recent reissued book, "Public Spaces in Japan" (Urban Design Movement, Shokokusha, See p. 37). The point argued is that we previously did not have public spaces in Japan but only those "Hiroba-ka" spaces.

In Western European urban structures with their Greek and Roman roots, public spaces, such as Agora and Imperial forums, were constructed in the middle of structures and they were used as markets for commercial transactions and the like, assembly halls for politics and trials, and stages for religious celebrations. Creating public spaces simply made the city.

In medieval times, communes surrounded by citadels arose and public spaces were constructed close to spiritual, political and economic centers such as churches, religious houses and town halls. In the Renaissance period and after the Baroque period, the logic of structuring whole cities was based around the spiritual core from which political and economic centers developed.

On the other hand, this diagram does not apply to the framework of traditional Japanese cities.

Fumihiko Maki applied his thought-provoking interpretation in his book, "The Obscured City" (Kajima

郭こそ河川や掘割、土塁などで防御しながら威風が整えられるものの、それは都市の「中心」というよりは象徴的な「奥」としての基点になるのであって、その周辺に武家町や商人町、職人町、寺町などが階層ごと職能ごとにゾーニングされながら計画的に配置されても、その街路構成は城に収斂しない。むしろ、筋違いや折れ曲がりといった迷路性が与えられて収斂することを拒んできた。

それは、城へ攻め込まれにくくするという軍事上の目的によるものだと一般にはいわれているが、一方で、それは「そうしたかった」からそういう形態をしているということも可能である。西洋の都市が、広場という中心に対して均質的に街路を区画するという構成を取りたがるのに対し、日本の都市は、奥へ向かって玉ねぎの皮のように多層に街区で取り囲む構成を指向する。それは民族性であり、根差す空間文化の違いである。

広場と参道

　伝統的な日本の都市では、その場ごとのロジックが優先され、全域を統一的なグリッド・パターンで構成するというような包括的論理があまりなじまなかったことはよく知られている。京都は、8世紀末に「平安京」として中国式の都市構成を参考に計画的に建設され、たしかに中心市街地には現在もその町割りは残っているが、その後市街がスプロールしてもそのロジックは延伸しなかった。やはり場所の論理が先行し、街路はねじれ、向きを変えていった。

　また、全国の坂や通りの名称に、富士見、汐見という類の名称は少なくないが、伝統的な都市で

Institute Publishing): Japanese urban structures as "Oku (deep inside) – Housetsu (subsumption)" as compared to Western European urban structures with ""Center – Block". If, in Western cities, "centers of the towns" are created with public spaces associated with core facilities and the town blocks are developed as they converge to the center, it is completely different from traditional Japanese urban structures.

In a castle town, the town is split partially into a grid pattern with the castle being at the base of the grid. A fortified castle, or citadel, can have dignity while being protected by a river, canal, earthworks and the like, and it is a symbolic "Oku" or "inside" of the greater city rather than the "center" of the city. The greater city may comprise several towns such as a samurai town, a merchant town, an artisan town and a temple town, be systematically zoned by class and occupation and be located around a castle, however, the street composition would not converge towards the castle. Rather, the towns were designed with diagonal or bending streets providing a maze-like structure and making convergence impossible.

Generally, it is believed that this maze-like structure served a military purpose, but it is possible that the city was formed in this way because people simply wanted it to be this way. Western cities tend to have a structure in which streets are divided homogeneously towards central public spaces. On the contrary, Japanese cities prefer a structure comprised of town blocks enveloped in multilayers like onion skin. This is deeply rooted in the Japanese cultural view of space.

Public Spaces and Sando (an approach to a shrine)

It is well known that comprehensive logic of structuring the entire area in a uniform grid-pattern was not adapted in traditional cities in Japan, but ad-hoc logic that was specific to each place was given priority instead. Kyoto was systematically constructed as "Heian-kyo" (ancient Kyoto) at the end of the 8th century with reference to the Chinese style of urban structure. Indeed, the current city center still has this structure, but as the city sprawled, this Chinese style was not applied to the existing areas. After all, the existing Japanese style logic was antecedent, and

は町割りを、しばしば遠望する山岳景や、海や湖沼といった水景に向けるということをしてきた。実に現場主義的な構成論理である。

と同時に、遠望する焦点に向けて街路軸を振るという意図には「奥性」が透けて見える。遠景に焦点を得ることで、その街路に奥行きが生まれる。それは西洋の広場が教会や市庁舎に取り付くのとは少々ニュアンスが違うものだ。

ロラン・バルトは、皇居を森に囲まれた空虚といったが、確かに日本の空間構造は、焦点に図像的クライマックスを持たない。周辺空間に奥性という方向性のみを演出し、アクセスするプロセスそのものが重視される傾向にある。

日本の伝統空間にこの構造は通底し、その典型的なものが「参道」である。

神社、寺院という日本の聖地は、かならず参道というアプローチ空間をもつ。逆にいうと西洋の宗教施設は、広場を必要としても、かならずしも参道は求めない。

日本の参道空間は、鳥居や橋あるいは山門といった、"結界"をくぐりつつ、奥へと誘う運動性をもった"みちゆき"空間であり、その焦点は空虚である。

仏教寺院では最奥に本堂（金堂）があり、それに向けて参道が配置される。しかし、本堂に到達するも、伽藍の奥は暗がりとなって見えず、その手前には香炉も炊き込まれて風景は霞んでいる。本堂の内部に入っても、ご本尊が安置されていることは分かっていても、通常はそれを見ることはできない（一定期間「御開帳」されることはある）。

神社も同様で、橋を渡り、鳥居をくぐって参道を進みゆき、ようやく終着点に到達するも、手前

the streets were already twisted and changed directions.

Names indicating "view", such as Fujimi (viewing Mt. Fuji) and Shiomi (viewing ocean) are used for lots of slopes and streets. In traditional cities, sometimes towns were spilt by taking a broad view of a mountain landscape or a water landscape (e.g. ocean or lake) into consideration. The structure logic is truly ad hoc or site specific.

At the same time, characteristics of Oku can be the intention of assigning street axes focusing on a broad view. That is, the streets gain depth by focusing on a broad view. It is a different nuance from Western public spaces which focus on a central church or town hall.

Roland Barthes said that the Imperial Palace is an empty space surrounded by forest. Admittedly, space structures in Japan do not have an iconic climax to focus on. Surrounding spaces produce only a perception of Oku and the process of access itself tends to be considered just as important.

This structure is connected with Japanese traditional spaces at a fundamental level and the typical example is "Sando" or an approach to a shrine.

Sacred places in Japan, shrines and temples, always have an approach space called "Sando". On the contrary, Western religious facilities require public space but not always an approach.

Japanese Sando spaces are Michiyuki or "travelling" spaces with motility, which tempt us towards the furthest depths while we are going through or under "Kekkai" (a barrier between the sacred and the secular), such as the "Torii" (shrine archway), bridge and temple gate, and its focus is empty space.

In a Buddhist temple, the main hall (Hondo or Kondo) is at the far end of Sando is directed towards the main hall. However, when we reach the main hall, we cannot see furthest depths of the temple and the view is hazy because of burning incenses in front of the temple. Even when we enter the inside of the main hall, the principal image or statue is most often not visible even though we know that it is always there. It is unveiled only at certain times (such occasion is called "Gokaicho").

The structure of shrines is the same as that of temples: we cross a bridge, go under Torii through Sando before we finally reach the destination. Although a magnificent shrine is in front of us, the main shrine is sur-

に豪壮な拝殿こそあれ、本殿は中垣に囲われ、入口にはひらりと白布が垂れ流されて奥は見えず、虚空の空間で完結する。あるいは本殿自体が存在せず、背後の山そのものがそれに相当するということもある。

要するに、極点が極点としての重力を持っていないのだ。

むしろ、焦点が曖昧だからこそ、プロセスの意味性が強まるとも考えられる。門前町や鳥居前町の側から見れば、その先に寺院の金堂や神社の社殿があることを「知って」いるが、見えない。しかも参道は、それ自体しばしば屈曲し、あるいは山門に大提灯が架けられて視界を遮る。むしろそれは「見え隠れ」を生み、その継起的なシークェンスの連続性こそが、奥へと人を導く運動性のエンジンとなるのだ。

日本の庭園様式を代表する、回遊式庭園も同様である。

茶道の台頭によって考案された、回遊式庭園もまた「見え隠れ」といって、わざと全貌を見せず一部を隠しながら（「障（さわ）り」という造園技法で）奥へ奥へといざなう運動性を持つ。桂離宮や修学院離宮など数々の名庭の構成原理である。

この運動性は、三次元的な空間を移動する中で、さまざまな意味やニュアンスを感じ取りながら、いつの間にか心象風景が体内に形成されることから生じる。伊藤ていじによれば「シンボルの分布という形で空間化が行われている」ということになる（『日本デザイン論』鹿島出版会）。そして、シンボルに挟まれた余韻、余白、空間が「間」であり、この概念こそ日本の伝統空間のキーワードである。

rounded by a middle fence and we cannot see its furthest depths because of a white cloth hanging down, fluttering and obscuring the view. It concludes with empty space. It also can be the case that a main shrine does not exist at all and instead a mountain serves as the equivalent.

The bottom line is that an extreme point does not have gravity as an extreme point.

Rather, characteristics of process can have greater meaning because the focus is unclear. When we are in Monzenmachi (temple town) or Toriimaemachi (town originally built around a shrine), we "know" that the Kondo of the temple or the main building of the shrine is there somewhere in front of us, but we cannot see them. Moreover, Sando sometimes curve or a large lantern is hung on the temple gate obscuring the view. It produces an "appearing and disappearing" condition, a continual sequence which becomes a motility engine to lead people to the far depths.

This idea can also apply to the circuit style garden, which is one of the main Japanese garden styles.

The circuit style garden, which was developed by the rise of the tea ceremony, also has the "appearing and disappearing" condition to tempt us to go further into the depths by not showing the whole garden at once and hiding a part of it (with a landscape gardening technique called "Sawari" (obstructing)). It is a design principle of many famous gardens, such as Katsura Imperial Villa and Shugakuin Imperial Villa.

This motility comes out when unnoticed imagined scenery is formed in our mind while we perceive various meanings and nuances as we travel through three-dimensional space. According to Teiji Ito, "Spatialization is performed in the shape of symbol distribution" ("Japanese Design Theory", Kajima Institute Publishing). And then, as an aftereffect, blank space and space between symbols are defined as "Ma" or "space in between" and this is a key concept of Japanese traditional space.

Structure of "Ma" (space in between)

One day, I was sorting out my photos. Soon after I looked over the slides of gardens and parks in Europe, I picked a slide of Katsura Imperial Villa up and was perplexed. It looked like a rambling mixed-up view for a moment.

「間」の構造

あるとき写真を整理していて、ヨーロッパの庭園と公園のスライドを眺め渡したすぐ後に、ふいに桂離宮のスライドを手にして戸惑ったことがあった。とりとめない、混乱した風景に一瞬見えたのだ。桂離宮のさまざまなシーンを被写体に収めようと思うとき、じつは絵画的構図と感じていたものが、物理的にはただの余白や空間である場所に、なんらかの心象風景を重ねあわせ、見る方が勝手に脳内でバランスさせて眺めていたことをそのとき知った。

「間」とは、心象に投影された空間的ヴォリュームである。前述の伊藤ていじはそれを「イマジナリー・スペース」と呼んだが、象徴的シンボルに囲まれて形成されるイマジナリー・スペースでは、その空間の調和やバランスの概念も、絶対座標軸を持ったユークリッド幾何学的世界観では説明し得ない。

西欧的概念の調和でいうハーモニー harmony とは、要素が抽象化され元素化した上で組織的に組み合わされるバランス状態である。クラシック音楽などの洋楽をイメージしてもらいたい。空間および時間は唯一つの視点（総括的展望）によって眺められる。いわば静的バランスといっていい。

これに対して、動的バランスによる調和概念をヘテロフォニー heterophony と呼ぶ。雅楽やアジアの民族音楽でこの用語は使われている。各音は抽象化というより象徴化であり、元素化されないまま何らかの濁りを保持した音で、さらに演者が任意で別々に動いたり、リズムやテンポを微妙に

We try to capture an object with lots of scenes of Katsura Imperial Villa as a pictorial composition. However, I learned that, in fact, we place some kind of landscape scenery upon a simple blank space or physical space and we balance them in our brain freely.

"Ma" is a spatial volume which is projected in the mind. Teiji Ito, who was mentioned above, called it "Imaginary Space". The Imaginary Space is formed by those symbolic symbols which surround it and it is impossible to explain its concept of spatial harmony and balance with the Euclidean geometry world view with absolute coordinate axes.

Harmony as a Western European concept is a balance condition in which elements are abstracted, transformed into the most basic form, and then combined systematically. Imagine Western music, such as classical music. Space and time can be viewed from only one viewing point (overall perspective). It is, as we say, static balance.

In contrast, a harmony concept with dynamic balance is called heterophony. This word is used in Japanese court music and Asian folk music. Each tone is symbolic rather than abstract and keeps some kind of dull sound without being transformed into the most basic form. Also, "Ma" or pause is produced by performers taking different actions as they choose, or changing the rhythm or timing overlays to create complex tunes with an improvised or instantaneous characteristic. In Western music, jazz, bossa nova derived from jazz and folk music, such as Portuguese fad and Spanish flamenco have similar characteristics. The common points are that the overall perspective is tenuous and successive sequence is considered as important.

Observers and listeners project their mental pictures onto the dull sound of the tune in the subject and "Ma" or pause between the beats. In fact, we experience the symbolic meanings between things through tune or objects while discovering these meanings by projecting onto our mental pictures. We can say that the roots are same: finding emotions from slight movement of Noh-mask and cheered up by a pose of Kabuki.

Returning to the topic of gardens, in contrast to the Western European geometric garden which has an overall focus of a fixed point, the Japanese garden is a space

ずらしたりすることで生じる「間」が、偶発性や瞬発性のある複雑な音色を重ね合わせる。洋楽でも、ジャズやそれから派生したボサノヴァ、あるいはポルトガルのファドやスペインのフラメンコなどといった民族音楽はこれに近い。共通するのは、総括的展望は希薄で、継起的連鎖が重視される点だ。

見る者、聞く者は、対象にある音の濁りやリズム的な「間」に心象を投影する。音や物を通して、実はモノとモノの間に象徴化された意味を心象投影的に見出しながら体験しているのである。能面のわずかな動きに喜怒哀楽を見るのも、歌舞伎の見得に気分を乗せるのも根は同じといっていい。

庭園に戻れば、定点という総括的展望を持つ西欧式幾何学庭園に対し、日本庭園は見る者が動き、あるいは刻々と変化する時間の中で心象風景と重ね合わせつつ体験する空間ということができる。ヴェルサイユ宮殿やヴォー・ル・ヴィコント宮に見られる、ル・ノートル式のパースペクティヴ空間と日本の回遊式庭園を比べればその違いは瞭然としている。

回遊式でなくとも、たとえば枯山水の傑作・龍安寺石庭(りょうあんじ)は、見る主体が動かなければどこに座っても決してすべての石を視野に入れることができない。各要素は海岸景観に見立てられたり、生き物の姿だといわれたりする象徴的造形である。これもまた動的バランスであり、景石は心象的な均衡としてバランスよく配置されたものなのだ。

日本の空間文化における「広場」を質的に考えるとき、そういう「間」の概念がもたらす、プロセス型の運動空間をプロトタイプの一つとして考えざるを得ない。それを参道的な奥行きを持った

where observers move or experience while overlaying imagined scenery in ever-changing time. If you compare the perspective space of Le Nôtre style garden, such as Palace and Park of Versailles and Chateau de Vaux-le-Vicomte, to a Japanese circuit style garden, the difference is obvious.

At Ryoanji, which is a non-circuit style garden and a masterpiece of Karesansui (dry landscape garden), not all the rocks are in view unless the observers themselves move along. It is a symbolic shaping in which each element is used to resemble a seashore landscape or a figure of a creature. This also has a dynamic balance and the stones are placed in the right equilibrium as a balance of mental pictures.

When we think qualitatively about "Hiroba" in Japanese space culture, we cannot avoid thinking that a process type of dynamic space is one of the prototypes provided by this "Ma" concept. By conceptualizing this prototype as a Michiyuki space with depth, it becomes easier to explain that the lively atmosphere, which would be observed in public spaces in Western Europe, can instead be seen on the streets within Japanese cities.

Now, is it not possible to shape the place with the lively atmosphere, which you can see in Western European public spaces, by being aware of the prototype and designing it intentionally? With this concept, as an urban designer, I designed a front approach to Izumo Oyashiro, "Shinmon-dori Street", which has just been completed in 2014.

Shinmon-dori Street as a Sando- type Public Space

Shinmon-dori Street (Fig. 1) is a relatively new Sando, which was built as a public street in 1913. The appearance of the Sando was completed in 1915: the large Torii was created at the foot of the Horikawa River with a dedication of Tokuichiro Kobayashi, who was a local person of distinction and a row of 280 pine trees were planted. At this time, the Sando was named as "Shinmon-dori" and a front approach of Izumo Oyashiro generally indicates this street now.

The essential points of Shinmon-dori Street are to have a more suitable appearance as Sando with a stone

「みちゆき」の空間として概念化すると、日本の都市で広場的なにぎわいが街路を中心に見られるということが説明しやすくなる。

では、そのプロトタイプを意識し、意図的にデザインすることで、広場的なにぎわいの場を造形することも可能ではないか。都市設計家としてそのようなコンセプトでデザインしたのが、2014年に完成したばかりの出雲大社表参道「神門通り」である。

参道型広場としての神門通り

神門通り（図1）は、大正2年（1913年）に道路として整えられた比較的新しい参道だ。大正4年に地元の名士である小林徳一郎氏の寄進により堀川のたもとに大鳥居が建てられ、併せて280本の松並木が植栽されて参道の体裁が整った。このとき「神門通り」と命名されて、今や出雲大社の表参道というとこの通りを指すのが一般となった。

この参道を石畳によってより参道らしい体裁に整え、なおかつ歩行者を優先させるよう造形化したのが神門通りの骨子なのだが、それを詳細に説明する余裕が本文にはない。

ここでは、「みちゆき」型のにぎわい空間の造形として、いわば広場的な機能を街路に実現を試みた主要部についてのみ解説する。

出雲大社の大鳥居が建つ、エントランス部周辺の交差点は、「勢溜（せいだまり）」と呼ばれ、この交差点は常に参詣人で混み合い、車が渋滞していた。神門通りは勢溜の直前で坂道になって駆け上がる。島根県は、狭かったその坂道部と勢溜交差点を、参拝客の規模に応じて改善すべく、沿道を用地買収し

path and to change the shape to prioritize pedestrians. However, there is no room to explain the details in this article.

In this article, I would like to explain the main point; to shape a street to give it the so-called public space function so that it creates a "Michiyuki" type of lively atmosphere.

The crossing around entrance area of Izumo Oyashiro, where the large Torii is, is called "Seidamari" and this crossing was always crowded with visitors and traffic jams. Shinmon-dori Street has a slope and it goes up just in front of Seidamari. In order to improve the narrow sloped part and the Seidamari crossing based on the extent of visitors, Shimane prefecture purchased an area along the approach and the width of that area was expanded. As a result, buildings along the approach of the slope part were rebuilt all together and the street space gained more design flexibility.

I designed the sloped part as a mixed space of slope and stairs.

The middle part of the slope became a combination of a sloped roadway and service road, and a sub line of flow which comprised stairs and flat spaces. Also, retaining walls with pines in squares and streetlights were intergraded into the boundary areas. Pedestrians can freely pass the sloped areas, stairs and flat spaces. The planting squares have a rounded shape along intertwined lines of flow and provide seating areas.

Thus, the sloped part of Shinmon-dori Street was shaped as a three-dimensional space.

This is a design which induces communication casually but also intentionally. The aim of this design was that the street would be used as public space with the

図1
神門通り坂道部。沿道側に階段と平場の組み合わされたサブ動線が設けられ、スロープ部の動線と絡み合う。境界部の植栽桝や擁壁は、ベンチや照明なども組み合わされた多目的な装置である。

Figure 1
The slope part of the Shinmon-dori Street. A sub line of flow which is a combination of stairs and flat spaces along the street is intertwined with a line of flow at a slope. Planting squares and retaining walls in the boundary area are multipurpose devices that benches, lights and the like are also combined into.

て幅員を大きく広げた。その結果、坂道部は沿道建物すべてが一斉に建て替わり、街路空間はデザインの自由度が上がることになった。

その坂道部を、スロープと階段の複合空間としてデザインした。

中央を車道と側道のスロープとし、沿道側に階段と平場を組み合わせたサブ動線を確保して、かつそれらの境界部に松の植栽桝擁壁や街灯を組み入れた。歩行者は、スロープと階段・平場部を自由に行き来できる。植栽桝は絡み合う動線に沿って丸みを帯びた形状となっており、人が座る場所も提供する。

こうして神門通りの坂道部は、三次元的な空間造形となった。

これは、さりげなくも意図的にコミュニケーションを誘発するデザインである。限られた街路幅員の中で、足早に通り抜ける人、ゆっくり沿道の街並みをひやかしながら歩く人、ベンチ擁壁に腰掛け休む人など、さまざまなアクティビティが、それぞれ秩序をもって共存することで、街路が広場のように利用されることを狙った。

参道型のにぎわい空間であり、日本的な「みちゆき空間」の造形である。

神門通りが完成した結果、歩行者の通行量はそれまでの10倍以上となり、シャッターだらけだった商店街は見違えるような活況となっている。

展開する駅前広場

日本には広場はなかったといったが、近年この図式に変化が生じつつある。中でも注目すべきは「駅前広場」である。

orderly coexistence of activities by various people within the limited width of the street: people passing through briskly, waking slowly, commenting on the street view, and resting at the retaining wall benches.

This shape created the "lively atmosphere" typical of Japanese Michiyuki space.

After the completion of Shinmon-dori Street project, the amount of pedestrian traffic increased 10 fold or more and the shopping district, where many shops were previously closed, changed beyond recognition and is now booming.

Developing Station Squares

As I mentioned above, we did not previously have public spaces in Japan. However, this has started to change in recent years. "Station squares" (at train stations) are a noticeable example of this change.

The main function of a station is to provide a stopping point for pedestrians to get on and off and transfer trains. Therefore, a "station square" is basically a public space for transportation; not only in Japan but all over the world. The space in front of a station, which provides a public space for pedestrians more than for roadway space, is unusual.

However, with time changing, cities sprawl and railway stations, which used to be at outskirts, have been gradually been incorporated into inner urban areas. Stations naturally are a facility producing a large number of users. It thus has enough capacity to become a hub. Moreover, Japan National Railways was divided into 6 different privatized railway companies by areas in 1987 and commercial elements were added to the management forms on a large scale. As a result, in addition to the original transport function, railway stations were combined with facilities including commercial facilities, hotels, and business facilities, and the capacity for functions as a hub improved dramatically.

The railway stations with commercial functions have shown how much they can attract people and it has affected the commercial potential of the surrounding urban areas. As a result, concourse space inside the facilities has expanded and been transformed to accom-

駅の主要機能は、いうまでもなく歩行者の乗降と乗り換えである。したがって「駅前広場」というものは、日本に限らず世界的にも基本的に交通のための広場でしかない。歩行者のオープンスペースを車道空間以上にもつ駅前空間というのは稀だ。

　しかし時代が推移し、都市がスプロールして、それまで町はずれにあった鉄道駅は次第に市街地の内部へ取り込まれるようになった。もともと大量の利用者を生む施設である。拠点化する資質は十分持っている。さらに日本は、1987年に国有鉄道が6つの地域ごとに分割民営化され、運営形態に商業性が大きく加味された。その結果、鉄道駅は本来の交通機能に加えて、商業やホテル、業務施設などと複合化し、拠点性を飛躍的に拡充した。

　商業施設化した鉄道駅は、周辺市街地の商業ポテンシャルに打撃を与えるほどの集客力を見せることになった。その結果、施設内部のコンコース空間は拡大と同時に多目的化し、駅舎の外部にも、西欧の都市広場と近似したオープンスペースを構えるようになってきた。

　しかし、このように特殊な成長を遂げつつある日本の駅前広場だが、今のところその多くはイベントスペース的なものであり、いわば商業空間の延長としてのオープンスペースに過ぎない。人々が生きている実感をもつほどの実存的価値をもった空間といえるものは少ない。このまま駅前に広場が造り続けられても、シエナのカンポ広場（図2）のようなクウォリティで人々に喜びを与える空間に成長するかは疑問である。

　それでも、今の状況はまさに戦後の都市計画家

modate multipurpose activities simultaneously. Also, areas outside of station buildings have started to become public space similar to the Western European urban public space.

　Although station squares in Japan have been specially developed, many of them are only used for special events and they are effectively just an expansion of commercial spaces. There are not many public spaces with an existential value which people can actually feel wonderful to be alive. There is some doubt whether the station squares develop to have a quality to give people a joy like Piazza del Campo in Siena (Fig. 2) even if they keep being constructed at the current pace.

　Even so, the current situation provides an opportunity to produce urban spaces in Japan which have the same quality as Western style public spaces, which was exactly what Japanese postwar urban planners wished for but could not have. Station squares are public spaces and they should be opened up for cities even though they are accompanied by train stations. By adhering to this approach, it becomes a touchstone of urban shaping for next generation. The Himeji station north exit

図2
カンポ広場（トスカーナ州シエナ、イタリア）
Figure 2
Piazza del Campo in Siena, Tuscany, Itary

図3
「キャッスルガーデン」と命名された姫路駅北駅前広場のサンクンガーデン。吟味した自然素材で造形され、各施設は空間内部に奥行きを創り出すよう意図的に平行配置で構成されている。
Figure 3
Sunken garden of the Himeji station north exit station square, called "Castle Garden". Examined natural materials are used to shape the garden and each facility is placed in parallel to create depth in the space.

が望みつつもなし得てこなかった、西洋型の広場と同等の質を日本の都市空間に実現させる契機である。ただし、駅前広場は公共空間であり、駅に付随するとはいえ都市に開かれた場であるべきものだ。そこに解答を与えてこそ、次世代の都市造形の試金石となる。それを追求したのが、本書でも紹介されている姫路駅北駅前広場（74頁・図3）である。チームで設計したものだが、ランドスケープ担当としては思いを込めた。

プラザとしての姫路駅北駅前広場

　姫路駅は、日本では珍しくバロック的なヴィスタ景を持つ都市軸の起点である。

　姫路市の中心部は第二次世界大戦の際に空襲によって甚大な被害を受けたが、その復興事業によって、1955年に姫路城と姫路駅とを結ぶ全長約1km、幅員約50mの直線的街路「大手前通り」が完成した。世界遺産・姫路城をアイストップとする都市軸は、東洋と西洋が融合したような風情である。

　近年姫路市は、鉄道の高架化事業と共に駅舎の改築、及び周辺整備に取り組んできた。なかでも北口の駅前広場は、日本最初の「トランジットモール」、姫路城への新たな視点場「キャッスルビュー（眺望デッキ）」、地下街の再整備に合わせた「キャッスルガーデン（サンクンガーデン）」と地上広場である「芝生広場」という多彩なメニューで構成されている。

　交通機能を集約させ、駅と姫路城を歩行者空間でダイレクトに結びつけたことは重要である。車両のための場から人のための広場へと大きく変貌

station square (p. 74, Fig. 3), which will be mentioned in this article, is an example of pursuing this view. This public space was designed by my team but I applied my thoughts while being in charge of the landscaping.

Himeji Station North Exit Station Square as a Plaza

Himeji station is a starting point of urban axis with a baroque style vista view, which is unusual in Japan.

　The central part of Himeji was largely damaged by air raids in World War 2. In the reconstruction project, "Ootemae-dori" Street was constructed in 1955. This street is a rectilinear street with total length of approximately 1 km and width of approximately 50 m connecting Himeji Castle and Himeji station. The urban axis with World Heritage site Himeji Castle as an eye stop has the flavor of East and the West fusion.

　Himeji city has made efforts with an overhead railway project, renovation of station buildings and rearrangement of the surrounding areas in recent years. Particularly, the Himeji station north exit station square is formed by combining various elements: a Japanese "transit mall", "Castle View, (viewing deck)" as a new viewpoint field for Himeji Castle, "Castle Garden (sunken garden)" for redevelopment of an underground mall, and a "public space with grass" on the ground.

　It was important that that road traffic function was concentrated and the station and Himeji Castle was connected directly by pedestrian space. The space was changed from a place for cars to a place for people and the whole station square became an entrance of travelling space like an approach to Himeji Castle. It is like a "preview" of a movie called Himeji Castle. I wanted people to expect a view of Himeji Castle there, even though it does not need to be the view of the castle itself; because the real castle exists at the end of the approach.

　Although the Castle View is shaped in the style of Modernism, it reminds us of a castle gate because it was built with steel and large sections of timber. The public space with grass on the ground captures the lively atmosphere with a simple green carpet and brick pavement.

　Moreover, you could say that the Castle Garden next to these spaces is a "three-dimensional public space"

を遂げた形だが、この駅前広場全体が姫路城へ続く、参道的なみちゆき空間のエントランスといえるものとなった。いわば、姫路城という映画の「予告編」である。ここでは姫路城の景色を期待させたい。しかし決してそのものである必要はない。本物はその先にあるのだから。

キャッスルビューはモダニズムの造形ながら、城門を思わせる、鋼鉄（くろがね）と大断面の木材で築かれた。地上部の芝生広場は、シンプルな緑のカーペットと煉瓦舗装でにぎわいを受け止める。

そして、これらに隣接するキャッスルガーデンは、地上と地下街、駅ビルの動線を結節する"立体広場"といっていいものだ。外周の壁面は城の石垣を思わせる錆御影石（さびみかげいし）を、石垣風ではなく、あえて矩形の石材を平積みに使った。その空間内部は、デッキテラスを中心に、野面積（のづら）みの石垣壁やスロープ、せせらぎなどが組み合わされ、素材感の中に歴史的な趣が込められている。また、それらは基本的にすべて長手方向に平行する形で整えられており、これは、中に入った人が一たび動き出せば、風景がスライドするように動き出し、立ち止まれば即座に静止するという効果を狙ったもので、日本庭園の技法の一つである。さりげなく見え隠れをつくりながら、単体空間の中に「奥性」を創り出すものであり、空間に参加することで自らの存在が実感されるという演出である。

この立体的かつ複合的な駅前広場は今や、どこに立っても広場というにふさわしい光景を獲得した。ただ座って景色を眺めている人がとにかく多い。そして広場でくつろぎ、会話を楽しみ、食事し、子供と遊び……という光景が夕方を過ぎても続く。西欧の広場と同等の"質"が創出されたと

connecting ground level to underground and a line of flow from the station building. Yellowish granite, which remind us of stonewall of a castle, were used for an external wall. However, they were intentionally laid out flat and not used to construct stonewall. Inside the space, the Nozura-zumi (laid with mostly unprocessed stones) stonewalls, slopes and a little stream are combined and historical taste is put into the materials. These are also shaped basically parallel in a longitudinal direction, and by doing so, we can produce an effect which is one of the techniques of Japanese gardening: once people inside start moving, the view also starts moving like a slide, and once people stop, the view also stops immediately. This technique produces "depth" in a single space while producing the condition of appearing and disappearing subtly. Also, this is a representation that we feel our existence by joining in the space.

This three-dimensional and composite station square acquires the appropriate scenery as a public space from any direction. Many people are just sitting and looking at the scenery. Even after evening, we can still see such views: people relaxing in the public space, enjoying their conversation, having a meal, playing with kids, etc. I could tell that the public space created the similar "quality" as that of Western European public spaces.

Future of "Hiroba" and Public Spaces in Japan

As I mentioned above, various station squares are in the process of gaining aspects of European plazas. Tokyo Station could become a center of railway stations in Japan, and a public space right in front of the station building at the Marunouchi Exit is going to be developed as a public space for people in the near future. Tokyo Station also has an urban axis viewing place to the Imperial Palace, which is called "Gyoko-dori" Street. Here there is a unique view of a vista on the opposite side of the station which then disappears into the Imperial Palace. The station square will have a plaza style public space in the center to catch this vista view and traffic will be dispersed and moved to the sides of the center public space.

On the other hand, even Shinmon-dori Street is unique as an "Sando-type public space", multiple examples in this article show that not only station squares

いう手応えのある空間となった。

日本の「広場」とパブリックスペースのこれから

さまざまな駅前広場がプラザ的な様相で実現しつつあると述べたが、日本の鉄道駅の中心ともいえる、東京駅の丸の内口も近年、その駅舎正面に人間のための広場が整備される予定である。東京駅は、その正面から皇居へと延びる都市軸「行幸通り」のアイストップでもあり、反対側は皇居へ消えるという独特のヴィスタの焦点である。駅前広場は、このヴィスタを受け止めるように、中央部分にプラザ型の広場を設け、交通広場はその両側に分散して配置されることになる。

一方、"参道型広場"としての神門通りは特殊だとしても、駅前広場以外にもさまざまな「広場」が、西欧広場とは異なるニュアンスで日本の都市景観を彩り始めたことは、本書のさまざまな事例が証明している。

日本の都市はこれまで広場という文化を持たなかった。

しかし、これからはさまざまな様相で広場的なオープンスペースが都市ごとに散りばめられる状況が期待される。それは確かに人間性の復権であり、伝統文化からの脱却というより、新たな展開であり発展といっていい。日本のパブリックスペースは、風土性と文化性を反映した独自の方法論で活性化しつつある。願わくはこのムーブメントが、日本の空間文化の新たな基盤へとつながるものであってほしい。

but also other various "Hiroba" have started to color the Japanese urban landscape with different nuances from Western European public spaces.

Japanese cities did not have a public space culture until now.

However, we can expect cities to be studded with plaza style public spaces in the future which will in turn provide a humanitarian restoration. Rather than growing out of traditional culture, it is a brand new development. We could even say it is an expansion. Public spaces in Japan are in the process of revitalizing Japan with a unique methodology reflecting the Japanese environment and cultural traits. Hopefully, this movement leads to a new foundation of Japanese space culture.

広場は"場""コンテンツ""スケール""時間""ネットワーク"5つの要素の組み合わせでできていると考えている。場があるだけでは広場にはならない。そこに目的があり、集う人がいてはじめて広場として成立する。そして集い方には広さと時間という物理的要因が関係してくる。現代の広場は5つの要素がフレキシブルに組み合わされ、目的のあり方、集い方そのものが多様化している。

広場化の歴史

日本の広場は昔から"広場化"がキーワードであった。神社仏閣の境内がある時には村人が集う祭事の場所となり、ある時は道が祝祭の場になるなど、普段は違う機能を持つ場所が、ある一定の時間に広場化し、機能する。広場化とは"コンテンツ"と"時間"によって"場"が変容することである。道を使う時にはいくつかの場所が同時期に"ネットワーク"化して使用されることも多い。そういったフレキシブルな広場のあり方は昔から日本人の精神性と通じる。物として固定化されたものよりも、常に変容し続ける"状態"の方を大切にしているからだ。

ネットワークによって生まれるプライベートな広場

"場"から用意されてしまった空間はあまり機能していないことが多い。公共建築の中につくられた広場。都市の至るところに残された空隙のような公開空地（そもそもが容積率、高さを稼ぐためなので目的が違うのだが）。むしろ"コンテン

In my opinion, public spaces are combinations of 5 elements being "place", "content", "scale", "time" and "network". Only having a place does not mean that it is a public space. Public space needs a purpose and must allow the gathering of people for the use of that place. The types of gatherings are determined by the factors of size and time. These 5 elements are combined flexibly in modern public spaces and the current purposes and gathering methods of public spaces are diverse.

History of "Hiroba-ka"

"Hiroba-ka" ("converting into an public space") has been a key word for public spaces in Japan. Sometimes the premises of shrines and temples are used for rituals where villagers gather and sometimes streets are used for celebrations. Places normally used for other purposes become public spaces for a certain period of time. Hiroba-ka means that a "place" transforms depending on the "content" and "time". Several places are often spontaneously used together in a "network" when streets

空間を超えた広場の在り方
永山祐子（永山祐子建築設計）

Hiroba beyond Space
Yuko Nagayama　(YUKO NAGAYAMA&ASSOCIATES)

ツ"が先にあり、それに合わせて発見された空間は魅力的だ。西荻窪南口からほど近い路地は幅3m程度の短い小道だけれど、夜になると道沿いのお店からテーブルと椅子が出され、道自体が集いの場になる。その小道に行けば顔なじみに会える。さらに最近はもっとプライベートな場が"広場化"するケースも見られる。誰かのリビングがライブ会場になり、その情報がFacebookなどソーシャルネットワークに載せられた瞬間、そこは公共の広場となる。広場となるのに大きさの制限はない。広場を不特定多数の人が自由に利用できる場と定義するならば、さらに話を広げられる。たとえば、ライドシェアリングと呼ばれる新しいシステムがある。自分が乗る車に空きがある時に、相乗り者をネットワーク上で募る。目的地と日程のあう人同士が車という小さな空間で集い、そこにコミュニケーションが生まれる。ネットワークにより、多様な目的によってさまざまな場に集うチャンスが増大した。誰でもその広場の管理者になれる。プライベートな場とパブリックの場の境界線は曖昧になり、一見すると分からない広場が街中に広がっている。

ネットワークによって広がるコンテンツ

おもちゃの交換会「かえっこバザール」というイベントが各地で開催されている。美術家である藤浩志氏が考案した仕組みである。子どもたちがいらないおもちゃを各自持ち寄り、イベント共通の通貨を発行し自由に買い物ができる。値段をつけるのも、売るのも、買うのも子どもたち。かえっこはおもちゃの交換だけが目的ではなく、

are used as public spaces. The state of public spaces is flexible and it always relates to Japanese spirituality: changing conditions are considered more important than fixed permanency for Japanese people.

Private Hiroba Created by Networking

A "space" deliberately created from "place" is often less functional. Hirobas created in public buildings and public Hiroba like voids are found everywhere in cities (they are for there, for example, for the sake of gaining floor-area ratio or height so their purpose is different in the first place). Rather, spaces which are found for their specific "content" are attractive. The alley near the south exit of the Nishiogikubo Station is a short path of approximately 3 m width. However, chairs and tables from shops along the path come out at nighttime and the path itself becomes a gathering space: you can see familiar faces there. Moreover, more private places are becoming "Hiroba-ka" nowadays. Someone's living room becomes a live site and as soon as the information is uploaded onto social network sites, such as Facebook, the place becomes a public space. The size of public spaces is not limited. If public space is defined as a place used by general public to their liking, we see that more places can be defined as public spaces. For example, rideshare is a new system: if you have spare seats in your car, you can look for someone on a social network to share the ride. People who have the same destination and matching dates gather in the small space of a car and communication between them starts. By using social networks, the opportunities increase for people to gather in various places for various purposes and anyone can be an administrator of the public space. The borderline of private place and public place becomes unclear and public spaces which cannot be distinguished at first glance are spread everywhere.

Contents Spread Out by Networking

"Kaekko" is an event where children can exchange their old toys and it has been around in various places. An artist, Hiroshi Fuji, developed the "Kaekko" system. In this event, children can get "Kaekko" money by bringing

売り買い行為そのものを楽しむ子ども主体のイベントである。そして藤浩志氏は余ったおもちゃを材料に彫刻を造っている。材料が生まれるプロセスから彼の作品となっているのである。かえっこは、時には1000人も集まることがあるほどに集客力があるので最近では防災イベントと組み合わせるなど、さまざまな目的と一緒に企画されることもあるという。「かえっこ」はその仕組みの詳細がネット上に公開されダウンロードできるので、誰でも意思さえあれば開催可能なため、日本全国はもちろんのこと、海外でも開催されるほどに普及している。"コンテンツ"は今や"ネットワーク"によって世界中どこでも簡単に手に入れることができ、それなりにノウハウの詰まったイベントが誰でも開催できるのである。そのような強力なコンテンツには人が集まる。アニメイベントなどの集客力を見ていると何かに特化したものであればあるほど集客力があがっている。特化することでターゲットが狭まると考えがちだが、その逆でかえってぴたりとはまるターゲットは増えている。これは今までパブリックな広場をつくるときの落とし穴でもあったような気がする。皆に良い場所をと平均化した結果、実は誰にも当てはまっていない中途半端なものができあがっているかもしれない。

民間企業内の公共スペース

会社の社屋の一部をパブリックな広場とした例がある。東京・芝浦にある「SHIBAURA HOUSE」(78頁)だ。創業当初は製版を主に請け負っていた会社が製版業をやめ、制作が主体と

their old toys and then use this money to buy other second hand toys. Children set the prices to sell and buy toys. "Kaekko" is an event for children and its purpose is not only exchanging old toys but also enjoying the experience of buying and selling. Moreover, Hiroshi Fuji creates sculptures using unsold toys and the origin of the materials forms a part of his artwork. Sometimes as many as 1000 people go to "Kaekko" events and other types of events are held in conjunction, such as events promoting disaster prevention. The details of the "Kaekko" system are publicly available on the web so anyone can download them and hold an event. Hence, it has been held not only in Japan but also overseas. "Content" can be obtained anywhere in the world through the use of a "network" and anyone can hold an event using this know-how. Such strong contents can attract people. By analysing the number of people attracted to events, such as animation events, it is determined that the more specific the events are, the greater the number of people will attend. We tend to think that the more specific an event is, the fewer people it will attract; on the contrary, the number of specifically targeted people actually increases for such events. This may be a pitfall in the process of creating public spaces. We try to provide a place which is balanced and appeals to everyone. But in the end, the place may end up unbalanced and suitable for no one.

Public Spaces in Private Company

"SHIBAURA HOUSE" (p.78) in Shibaura, Tokyo, opens spaces in its building for the public. At the beginning of its establishment, it undertook mainly plate-making but it stopped palate-making and focused on production. As a result, the large spaces in the building, where printing machines had been, were vacated and unused. Generally unused spaces in a building would be rented out as offices; however, these spaces in Shibara House were used as public spaces. Kazuyo Sejima designed the new office building and lots of spaces in the building became open spaces. The walls of the open spaces are glass and people can see inside from outside. These open spaces become an oasis for neighbors in Shibaura, where there are not many parks. In the morning, elderly neighbors

なったことで大きな場所を占めていた印刷機がなくなって屋内に場所が余る。その場所を、通常は賃貸オフィスにすると考えるところを公共の場としてオープンなスペースにしようと考えた。そして新しい社屋の設計を妹島和世氏に依頼し、新社屋内の多くの場所をオープンにした。ガラス貼りの外からも丸見えのオープンスペースは、公園が少ない芝浦界隈の人々の憩いの広場となっている。天井が高く明るい外部のような開放性を持った気持ちの良いスペースには午前中は近くのお年寄りが集まり、お昼すぎになると幼稚園帰りのお母さんと子どもたちが集まる。魅力的なイベントも多数企画され、さまざまな人が出入りすることで会社自体の活性化にもつながっている。ここまで公共に貢献している場でありながら自治体とは無関係に存在している。各自治体が税制の優遇措置などで民間の広場づくりの後押しをしたら、自治体が自力で場所を確保して一からつくり上げていくよりもはるかに少ない労力でしかもスピーディに、より魅力的で多様な広場が街のあちこちに増えてくるかもしれない。

木屋旅館（82頁）

愛媛県宇和島市にある明治44年創業の老舗旅館を、一日1客という新しい形態の旅館にリノベーションした。このプロジェクトは松山からの高速道の開通にあわせ、街を活性化しようと宇和島市と合同会社きさいや宇和島が共同で行った。宇和島には昔からお構の文化がある。いわゆる「頼母子講（たのもしこう）」という、昔から続く相互金融システムである。宇和島に住む多くの人がたくさんの

gather in the relaxing spaces with high ceilings and bright openness similar to an outside space, and in the afternoon, mothers and their children gather on the way back from kindergartens. A number of attractive events have been held and various people go in and out of the building; it leads to the revitalization of the company. Although these open spaces contribute to the public, the local governments have not been involved. If each local government promotes the development of private Hiroba with such measures as preferential taxation and the like, the number of more attractive and diverse public spaces may increase. Private Hiroba can be created faster and with less labor as compared to those created by local governments which have to first find suitable places and create the public space from scratch.

Kiya Ryokan

Kiya Ryokan (p. 82) is an old hotel opened in 1911 and it locates in Uwajima, Ehime. It was renovated and has a new system that only one group of people can stay for a night. Uwajima city and a local limited partnership company (KISAIYA UWAJIMA) set up the project together to revitalize the city when the expressway from Matsuyama was opened. "Okou" culture has been around in Uwajima. It is the so-called "tanomoshikou" which is a mutual financing system. Many people in Uwajima belong to Okou and they have regular meetings at restaurants. The restaurant business is revitalized because many Okou meetings, such as Okou for wives in day time and Okou for husbands at night, are held regularly. Kiya Ryokan also has a traditional tatami mat room where Okou can be held. By taking this locality into consideration in the design process, I believed that although Kiya Ryokan is an accommodation which is a private space, it should be open to the public for certain occasions. After completion, I thought about organizing something for the public at Kiya Ryokan, and an art event "AT ART UWAJIMA 2013" was held in 2013. An image artist, Tabaimo and a comic artist, Yoriko Hoshi collaborated in this event. "a tinge of KIYA" is an image work by Tabaimo based on a novel which is based in Kiya Ryokan and written by Yoriko Hoshi. It was projected in various places in the Ryokan and it was

お構に所属していて、会合を飲食店で定期的に行う。昼は奥様方のお構、夜は旦那衆のお構という具合に沢山の会合があるため、飲食業界は活性化している。木屋旅館でもお構ができるような座敷を用意した。そのような地域性もあって、設計にあたっては宿泊施設というプライベートな場ではあるけれど、時には街へ開かれた場になるようにしたいと考えた。完成後、宿泊施設だけではない場所として自分でも何か企画してみたいと思い立ち、2013年にアートイベントである「AT ART UWAJIMA2013」を企画した。映像アーティスト束芋氏と漫画家ほしよりこ氏をコラボレーターとして迎え、木屋旅館を舞台としたほしよりこ氏の小説をもとにした束芋氏の映像作品「木屋の染み」が旅館内のさまざまな場所に投影され、木屋旅館の空間と融合した新しい表現となった。オープニング時を含めて、宇和島市はもとより多方面から多くの人が見に訪れてくれた。アートというコンテンツによって古い旅館が開かれた場所となった。

「届かない場所」という広場の在り方

どんな小さな場所であっても誰もが広場化できる可能性がある。誰でも入っていけるフラットな広場が色々な所に増えていくことは街をより面白くすると思う。しかし、一方で広場の象徴性みたいなものは薄まっている気がする。象徴となることも広場の役割であった。年に一度だけ祝祭を行う境内。普段は誰も足を踏み入れられない場所。突然目の前に開けた何もない場所。それはそれで魅力を持っている。象徴の広場は、別に入れなく

a new representation of images fused with the spaces in the Ryokan. A number of people from Uwajima and also other areas visited Kiya Ryokan to see this image work including at the opening premier event. Thus an old Ryokan became a public space through "content", in this case art.

Public Space as "Unreachable place"

Even very small places can be turned into Hiroba or a public space by anyone. I believe that increasing available flat public spaces in various places, where anyone can enter, makes a town more interesting. However, on the other hand, some kind of symbolism of public spaces may have been weakned. The other role of public spaces was to become a symbol: premises where a celebration is held only once a year, a place where normally no one can enter, and a vacant place which suddenly opens up in front us. These places have their own attractiveness. You do not need to enter the symbolic public spaces. The existence of the places can release our spirits. If the public spaces mentioned above are functional public spaces, the symbolic public spaces are for the sprits. This world becomes very convenient and we can even go to mysterious lands. There is almost nowhere we cannot go or do not know. I started thinking that unreachable places are important in my design at a certain point of my life. Although it is physically inconvenient that there is a place no one owns or no one can go to, conversely, I believe that spiritually it is free and rich. We think about the place in front of us, where we feel we can reach but cannot go. In that moment, I feel like my spirit is released from my body. An example of such a place is a garden in a Zen temple: the most famous garden in Ryoanji. People cannot enter the garden but they sit in front of the garden and throw their spirits into the empty space by looking at it. However, it does not mean that all the empty spaces work in this way. The place has to be attractive enough to throw our spirits into.

We have been thinking about various open places, but how are attractive public spaces, such as attractive voids and places with multiple possibilities, created? Public spaces are not solid but completely "empty" spaces. It

てもいい。そこにあるだけで精神を解放してくれる場所だ。先ほどまで話していた広場を機能としての広場とすれば、象徴の広場は精神にとっての広場になる。この世の中、秘境と呼ばれるところにも行けるほどに便利になり、誰も行けない場所、誰も知らない場所などもうほとんど残されていない。私はある時から自分の設計の中で"届かない場所"を大切に考えている。それは誰の所有でもない、誰も行けない場があることは物理的には不自由に感じるけれど、精神的には逆に自由で豊かなのではないかと思うからだ。手が届きそうだけれど容易には行けない場を前にして、そこに思いを馳せる。その瞬間、精神が身体を超えて解放されるように思う。たとえば禅寺のお庭がそうだ。最も有名な龍安寺の石庭。誰も入れないけれど、皆その前に座って眺めながらそのぽっかり空いた空間に精神を投影している。ただ間違ってはいけないのは、ぽっかり開けた空地であればいいというわけでは決してない。そこが精神を投影するに足る魅力的な地でなくてはいけない。

　さまざまな広場について考えてきたが、魅力的な広場とはどうつくられるのだろうか。魅力的な空隙。色々な可能性のある場所とは。広場とはあくまで実ではなく「空」の場である。広場を単体として考えるのは難しそうだ。広場を広場たらしめているものは、結局のところ取り巻く環境、状況である。魅力的なアクセス、魅力的なシチュエーション、魅力的なコンテンツ、そのすべてが揃った結果、そこに魅力的な広場がぽっかりと浮かび上がってくる。そう考えると広場とは、時代の映し鏡のようである。

probably is difficult to think about public spaces as an elemental substance. Making public spaces what they should be is, after all, contingent on the surrounding environments and conditions. The attractive public spaces will freely come to the fore when all the elements, which are attractive access, attractive situation and attractive content, are completed. By viewing them in this way, public spaces are like a mirror reflecting the times.

掲載事例 所在地一覧
Location list

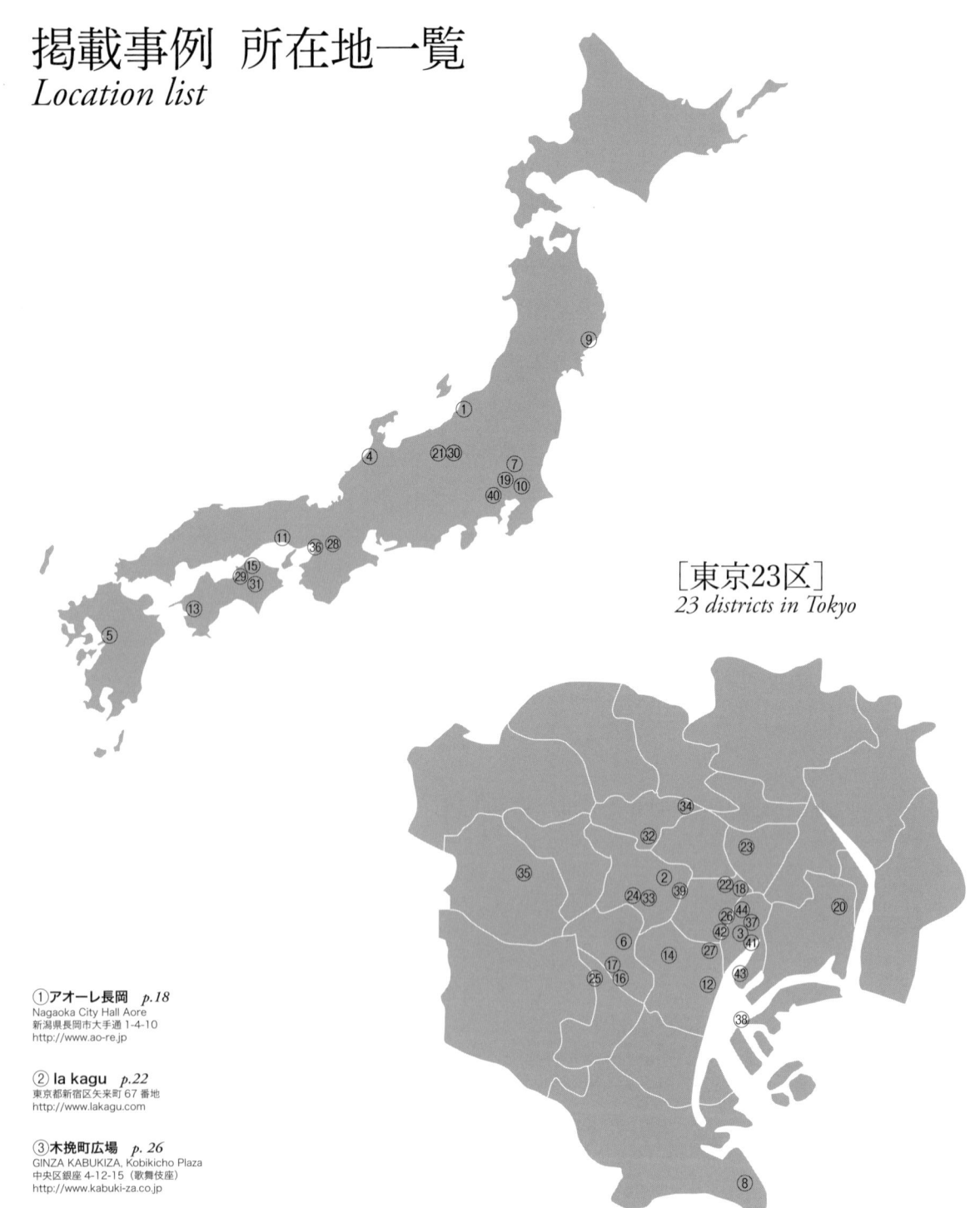

[東京23区]
23 districts in Tokyo

① アオーレ長岡 *p.18*
Nagaoka City Hall Aore
新潟県長岡市大手通 1-4-10
http://www.ao-re.jp

② la kagu *p.22*
東京都新宿区矢来町 67 番地
http://www.lakagu.com

③ 木挽町広場 *p. 26*
GINZA KABUKIZA, Kobikicho Plaza
中央区銀座 4-12-15（歌舞伎座）
http://www.kabuki-za.co.jp

④ 金沢 21 世紀美術館 *p. 54*
21st Century Museum of Contemporary Art, Kanazawa
石川県金沢市広坂 1-2-1
https://www.kanazawa21.jp

⑤ 熊本駅東口駅前広場（暫定形） *p. 56*
Kumamoto Station east exit station square
熊本県熊本市春日3- 637－2地先

⑥ 東急プラザ 表参道原宿「おもはらの森」 *p. 58*
TOKYU PLAZA OMOTESANDO HARAJUKU, "OMOHARA-NO-MORI"
東京都渋谷区神宮前 4-30-3
http://omohara.tokyu-plaza.com

⑦ 錄 museum *p. 62*
Roku museum&cafe
栃木県小山市東城南 2-23-5
www.roku-museum.com

⑧ 東京国際空港第2旅客ターミナルビル *p. 64*
Tokyo International Airport Terminal 2, "UPPER DECK TOKYO"
東京都大田区羽田空港

⑨ りくカフェ *p. 68*
Rikucafe
岩手県陸前高田市高田町字鳴石 22-9
http://rikucafe.jp

⑩ KOIL（柏の葉オープンイノベーションラボ）イノベーションフロア *p. 70*
Kashiwa-no-ha Open Innovation Lab., Innovation Floor
千葉県柏市若柴 178 番地 4　柏の葉キャンパス 148 街区 2　ショップ&オフィス棟6階
http://koil.jp

164

⑪姫路駅北駅前広場　p. 74
Himeji station North square
兵庫県姫路市駅前町 188（姫路駅）

⑫SHIBAURA HOUSE　p. 78
東京都港区芝浦 3-15-4
http://www.shibaurahouse.jp

⑬木屋旅館　p. 82
KIYA RYOKAN
愛媛県宇和島市本町追手 2-8-2
http://kiyaryokan.com

⑭六本木ヒルズ 66 プラザ　p. 84
Roppongi Hills 66 plaza
東京都港区六本木 6丁目 10−1 六本木ヒルズ 2F

⑮仏生山温泉（仏生山まちぐるみ旅館）　p. 86
Busshozan-onsen (Busshozan-machigurumi-hotel)
香川県高松市仏生山町乙 114-5
http://busshozan.com

⑯代官山ヒルサイドテラス　p. 90
HILLSIDE TERRACE
東京都渋谷区猿楽町 29-18
http://www.hillsideterrace.com

⑰代官山 蔦屋書店　p. 92
DAIKANYAMA T-SITE
東京都渋谷区猿楽町 17-5（蔦屋書店）
http://tsite.jp/daikanyama

⑱マーチエキュート神田万世橋　p. 94
mAAch ecute KANDA MANSEIBASHI
東京都千代田区神田須田町 1-25-4
http://www.maach-ecute.jp

⑲東北自動車道　羽生パーキングエリア（上り線）　p. 96
The Tohoku Expressway Hanyu Parking Area(Up Line)
埼玉県羽生市弥勒字五軒 1686
http://www.driveplaza.com/special/onihei

⑳サンストリート亀戸　p. 98
Sun Street KAMEIDO
東京都江東区亀戸 6-31-1
http://www.sunstreet.co.jp

㉑権堂パブリックスペース OPEN　p. 100
Gondo Public space OPEN
長野県長野市権堂町 2300
http://open-gondo.com

㉒アーツ千代田 3331　p. 102
3331 Arts Chiyoda
東京都千代田区外神田 6丁目 11-14
http://www.3331.jp/

㉓SCAI THE BATHHOUSE　p. 104
東京都台東区谷中 6-1-23
tp://www.scaithebathhouse.com

㉔新宿マルイ 本館　屋上庭園「Q-COURT」　p. 106
Shinjuku Marui Main Building, rooftop garden "Q-COURT"
東京都新宿区新宿 3-30-13
http://www.0101.co.jp

㉕目黒天空庭園　p. 108
Meguro Sky Garden
東京都目黒区大橋一丁目 9 番 2 号
https://www.city.meguro.tokyo.jp/shisetsu/shisetsu/koen/tenku.html

㉖丸の内ブリックスクエア　三菱一号館広場　p. 112
Marunouchi BRICK SQUARE
東京都千代田区丸の内 2-6-1 丸の内ブリックスクエア（パークビル）
http://www.marunouchi.com/top/bricksquare

㉗虎ノ門ヒルズ　オーバル広場　p. 114
Toranomon Hills OVAL PLAZA
東京都港区虎ノ門 1丁目（虎ノ門ヒルズ）
http://toranomonhills.com/ja

㉘興福寺　薪御能（薪能）　p. 116
The Kohfukuji Temple, Takigi-Onoh (Takigi-Noh)
奈良県奈良市登大路町 4−8

㉙旧金毘羅大芝居（金丸座）　p. 117
The Former Kompira Oshibai (Kanamaruza)
香川県仲多度郡琴平町乙 1241
http://www.konpirakabuki.jp

㉚門前町（善光寺）　p. 118
Cathedral town (Zenkoji)
長野県長野市元善町 491（善光寺）

㉛参道（金刀比羅宮）　p. 119
The approach to Shrine (Kotohira-gu)
香川県仲多度郡琴平町 892-1（金刀比羅宮）
http://www.konpira.or.jp

㉜鬼子母神　手創り市　p. 120
Kishimojin, Marche
東京都豊島区雑司が谷 3-15-20
http://www.kishimojin.jp

㉝花園神社　酉の市　p. 122
Hanazono Shrine, "Tori-no-ichi"
東京都新宿区新宿 5-17-3
http://www.hanazono-jinja.or.jp

㉞巣鴨地蔵通り商店街（「とげぬき地蔵」）　p. 124
Sugamo Jizo Street Mall ("Togenuki Jizo")
東京都豊島区巣鴨 3-35-2（高岩寺）
http://www.sugamo.or.jp

㉟阿佐ヶ谷駅前広場　p. 126
Asagaya station square
東杉並区阿佐谷南 3-36-2（阿佐ヶ谷駅）

㊱ポンテ広場　p. 128
PONTE SQUARE
大阪府大阪市浪速区湊町 1-4-1(OCAT)
http://ocat.co.jp/ponte

㊲歩行者天国（銀座）　p. 130
Pedestrian mall (Ginza)
東京都中央区銀座　中央通り

㊳お台場海浜公園　p. 132
Odaiba Marine Park
東京都港区台場 1 丁目 お台場海浜公園

㊴カナルカフェ　p. 134
CANAL CAFE
東京都新宿区神楽坂 1-9
http://www.canalcafe.jp

㊵お鷹の道　p. 136
Otaka-no-michi
東京都国分寺市西元町 3 丁目

㊶隅田川テラス　p. 137
Sumidagawa Terrace
東京都中央区明石町 8−1（聖路加ガーデン）

㊷ネオ屋台村® 有楽町東京国際フォーラム村　p. 138
NEO STALL VILLAGE®
東京都千代田区丸の内 3-5-1 東京国際フォーラム 1 階地上広場
http://www.w-tokyodo.com/neostall

㊸太陽のマルシェ　p. 140
Taiyo-no-marche
東京都中央区勝どき 1-9-8
http://timealive.jp

㊹橋詰広場（日本橋）　p. 142
Bridgehead plaza (Nihombashi)
東京都中央区日本橋

マルセイユ現代美術センター　p. 30
FRAC Marseille
フランス、マルセイユ

ブザンソン芸術文化センター　p. 32
Besançon Art Center and Cité de la Musique
フランス、ブザンソン

著者紹介
About Authors

隈 研吾　*Kengo Kuma*

1954年生。東京大学建築学科大学院修了。1990年、隈研吾建築都市設計事務所設立。現在、東京大学教授。1997年「森舞台／登米市伝統継承館」で日本建築学会賞受賞、その後「水／ガラス」(1995)、「石の美術館」(2000)「馬頭広重美術館」(2000) 等の作品に対し、海外からの受賞も数多い。2010年「根津美術館」で毎日芸術賞。近作に浅草文化観光センター (2012)、長岡市役所アオーレ (2012)、「歌舞伎座」(2013)、ブザンソン芸術文化センター(2013)、ＦＲＡＣマルセイユ(2013)等。著書に『自然な建築』(岩波新書)、『小さな建築』(岩波書店)、『日本人はどう死ぬべきか？』(養老孟司氏との共著　日経BP社)、『建築家、走る』(新潮社)、『僕の場所』(大和書房) などがある。

Born in 1954, completed Graduate School of Engineering (Department of Architecture), The University of Tokyo. Established Kengo Kuma & Associates in 1990. Currently a professor at The University of Tokyo. In 1997 he won the Architectural Institute of Japan Award for "Noh Stage in the Forest / Tome-City Traditional Performing Arts Inherit Hall." Later on, he won many prizes from overseas, for example, "Water/Glass (1995)," "Stone Museum (2000)," "Nakagawa-machi Bato Hiroshige Museum (2000)," etc. In 2010, he won the Mainichi Art Award for "Nezu Museum." The recent works include Asakusa Tourist Information Center (2012), Nagaoka City Hall Aore (2012), "Kabuki-za (2013)," Besançon Art Center and Cité de la Musique (2013), FRAC Marseille (2013), etc. His literacy works include "Shizen na Kenchiku (Iwanami shoten)," "Chisana Kenchiku (Iwanami shoten)," "Nihonjin wa dou sinubekika? (co-authored with Takeshi Yorou / Nikkei BP)," "Kenchikuka, Hashiru (Shincho-sha)," "Boku no Basho (Daiwa Shobo)," etc.

陣内秀信　*Hidenobu Jinnai*

1947年生。東京大学大学院工学系研究科博士課程修了。イタリア政府給費留学生としてヴェネツィア建築大学に留学、ユネスコのローマ・センターで研修。現在、法政大学デザイン工学部教授。専門はイタリア建築史・都市史。主な受賞歴に、サントリー学芸賞、地中海学会賞、イタリア共和国功労勲章（ウッフィチャーレ章）、パルマ「水の書物」国際賞、ローマ大学名誉学士号、サルデーニャ建築賞2008などがある。著書に、『東京の空間人類学』(筑摩書房)、『ヴェネツィア‐水上の迷宮都市』(講談社)、『都市と人間』(岩波書店)、『シチリア‐＜南＞の再発見』(淡交社)、『地中海世界の都市と住居』(山川出版社)、『イタリア　小さなまちの底力』(講談社)、『迷宮都市ヴェネツィアを歩く』(角川書店)、『イタリア海洋都市の精神』(講談社)、『イタリアの街角から－スローシティを歩く』(弦書房)、『アンダルシアの都市と田園』(編著・鹿島出版会)、『水の都市　江戸・東京』(編著・講談社) などがある。

Born in 1947, completed Doctor's Course at Graduate School of Engineering, The University of Tokyo. Studied at University Iuav of Venice through the Italian Government Scholarship. Trained at Historic Centre of Rome by UNESCO. Currently a professor at Faculty of Engineering and Design, Hosei University. Specializes in the Italian architectural history and urban history. Main awards include Suntory Prize for Social Sciences and Humanities, Collegium Mediterranistarum Award, Ordine al merito della Repubblica Italiana (ufficiale), Parma "Book of Water" International Award, University of Roma Honorary Bachelor's Degree, Sardegna Architectural Award 2008, etc. His literacy works include "Tokyo no Kukan Jinruigaku (Chikuma Shobo)," "Venezia – Suijo no Meikyu Toshi (Kodansha)," "Toshi to Ningen (Iwanami Shoten)," "Sicilia -<Minami> no Saihakken (Tankosha)," "Chichikai Sekai no Toshi to Jukyo (Yamakawa Shuppansha)," "Italia –Chisana Machi no Sokojikara (Kodansha)," "Meikyu Toshi Venezia wo Aruku (Kadokawa Shoten)," "Italia Kaiyo Toshi no Seishin (Kodansha)," "Italia no Machikado kara Slow City wo Aruku (Genshobo)," "Andalusia no Toshi to Denen (author and editor / Kajima Institute Publishing)," "Mizu no Toshi –Edo, Tokyo (author and editor / Kodansha)," etc.

小野寺康　　Yasushi Onodera

1962年生。東京工業大学大学院社会工学専攻修士課程修了。アプル総合計画事務所を経て、1993年小野寺康都市設計事務所設立。主な受賞歴に、土木学会デザイン賞・最優秀賞：「門司港レトロ地区環境整備」（2001年）、「津和野 本町・祇園丁通り」（2009年）、「油津堀川運河整備事業」（2010年）、「日向市駅及び駅前周辺地区デザイン」（2014年）BCS賞：「日向市駅駅前広場」（2009年）。主な著書に、『広場のデザイン－「にぎわい」の都市設計5原則』（彰国社）、『GS群団奮闘記 都市の水辺をデザインする』（共著、彰国社）などがある。

Born in 1962, completed Master's Course at Department of Social Engineering, Graduate School of Tokyo Institute of Technology. After working at APL Associates Inc., established Office of Yasushi ONODERA –Civil Engineering & Landscape Architecture Design Office- in 1993. Main awards include JSCE Civil Engineering Design Prize, Grand Prize: "Urban Landscape Design Project in Mojikoh area (2001)," "Honcho & Gioncho Dori Street in Tsuwano (2009)," "Horikawa-Canal in Aburatsu (2010)," "Hyugashi Station and the station plaza district (2014)," BCS Award: "Hyugashi Station Plaza (2009)," etc. Main literacy works include "Hiroba no Design (Shokokusha)," "GS Gundan Funtoki -Toshi no Mizube wo Design Suru (co-authored / Shokokusha)," etc.

永山祐子　　Yuko Nagayama

1975年生。1998年昭和女子大学生活美学科卒業。1998〜2002年、青木淳建築計画事務所勤務。2002年永山祐子建築設計設立。主な仕事に「LOUIS VUITTON 京都大丸店」、「丘のある家」「ANTEPRIMA Singapore ION店」、「カヤバ珈琲」、「SISII PRESSROOM」「木屋旅館」「豊島横尾館」など。主な受賞歴に、ロレアル賞奨励賞、JCDデザイン賞奨励賞、AR Awards（UK）優秀賞「丘のあるいえ」、（2006）、ARCHITECTURAL RECORD Award, Design Vanguard2012、JIA新人賞「豊島横尾館」（2015）などがある。
http://www.yukonagayama.co.jp

Born in 1975, completed undergraduate course at Faculty of Human Life and Environmental Sciences, Showa Women's University in 1998. Worked at Jun Aoki & Associates from 1998 to 2002. Established Yuko Nagayama & Associates in 2002. Main works include "LOUIS VUITTON kyoto daimaru, " "a hill on a house," "ANTEPRIMA SINGAPORE ION," "Kayaba Coffee," "sisii Pressroom," "KIYA RYOKAN," "Teshima Yokoo House," etc. Main awards include LO'real Color, science and arts Award Encouragement Prize, JCD Design Award 2005, AR Awards Highly commended prize (UK) on "a hill on a house (2006)," ARCHITECTURAL RECORD Award, Design Vanguard 2012, JIA Young Architect Award on "Teshima Yokoo House" (2015), etc. http://www.yukonagayama.co.jp

鈴木知之　　Tomoyuki Suzuki

1963年生。明治大学建築学科卒業、東京都立大学大学院建築学科修士課程修了。山本理顕設計工場勤務後、都市・建築写真家に転向。法政大学エコ研兼任研究員として、南イタリア調査／陣内研究室（2008〜2009）、東北調査／岡本研究室（2013）に参加。現代写真研究所講師。主な活動に、個展「Roji」／コニカプラザ・新宿（2001）、個展「Parallelismo」／リコー RINGCUBE・銀座（2011）、雑誌『東京人』（2002〜）、雑誌『NATIONAL GEOGRAPHIC 日本版』10月号（2012）などがある。

Born in 1963, completed undergraduate course at Department of Architecture, Meiji University. Completed master's course at Department of Architecture, Tokyo City University, After working at Riken Yamamoto & FIELDSHOP, became a city/ architecture photographer. Participated in a field study in south Italy / Jinnai Laboratory (2008-2009), and a field study in Tohoku Region/ Okamoto Laboratory (2013), as an additional post researcher from Laboratory of Regional Design with Ecology, Hosei University. Lecturer at Contemporary Photography Research Institute. Main activities include a solo exhibition "Roji / Konika Minolta Plaza, Shinjuku (2001)," solo exhibition "Parallelismo / Ricoh RING CUBE, Ginza (2011)," magazine "Tokyo Jin (2002-)," magazine "NATIONAL GEOGRAPHIC Japan (October 2012)," etc.

翻訳
Translation
pp. 6-14　　アルフレッド・バーンバウム　　Alfred Birnbaum
Other pages　　牧尾晴喜（株式会社フレーズクレーズ）　　Haruki Makio (Fraze Craze Inc.)

ブックデザイン
Book Design

鈴木正道　　*Masamichi Suzuki (Suzuki Design)*

本書の制作にあたり、多大なご協力を賜りました諸機関、関係者各位に厚く御礼申し上げます。
We would like to express our sincere gratitude to all those involved for their generous contributions toward the production of this book.

広場
Hiroba: All about "Public spaces" in Japan

平成 27 年 4 月 4 日　　初版発行

監修　　隈研吾　陣内秀信
写真　　鈴木知之
発行者　納屋嘉人
発行所　株式会社 淡交社
　　　　本社　〒603-8588　京都市北区堀川通鞍馬口上ル
　　　　営業　075-432-5151
　　　　編集　075-432-5161
　　　　支社　〒162-0061　東京都新宿区市谷柳町 39-1
　　　　営業　03-5269-7941
　　　　編集　03-5269-1691
http://www.tankosha.co.jp
印刷・製本　図書印刷株式会社

ⓒ2015　淡交社　　Printed in Japan
ISBN978-4-473-04014-5

落丁・乱丁本がございましたら、小社「出版営業部」宛にお送りください。
送料小社負担にてお取り替えいたします。
本書の無断複写は、著作権法上での例外を除き、禁じられています。